AF360920

Mobile Diagnosis 2.0

Mobile Diagnosis 2.0

Editors

Aniruddha Ray
Hatice Ceylan Koydemir

MDPI • Basel • Beijing • Wuhan • Barcelona • Belgrade • Manchester • Tokyo • Cluj • Tianjin

Editors
Aniruddha Ray
Physics and Astronomy,
University of Toledo
USA

Hatice Ceylan Koydemir
Electrical and Computer Engineering,
University of California
USA

Editorial Office
MDPI
St. Alban-Anlage 66
4052 Basel, Switzerland

This is a reprint of articles from the Special Issue published online in the open access journal *Diagnostics* (ISSN 2075-4418) (available at: https://www.mdpi.com/journal/diagnostics/special_issues/mobile_diagnosis_2).

For citation purposes, cite each article independently as indicated on the article page online and as indicated below:

LastName, A.A.; LastName, B.B.; LastName, C.C. Article Title. *Journal Name* **Year**, *Volume Number*, Page Range.

ISBN 978-3-0365-0380-6 (Hbk)
ISBN 978-3-0365-0381-3 (PDF)

Cover image courtesy of Hatice Ceylan Koydemir.

Contents

About the Editors

Aniruddha Ray is an Assistant Professor in the Department of Physics and Astronomy at the University of Toledo, Ohio, USA. He completed his Bachelors and Master's in Physics at the Indian Institute of Technology, Kharagpur, India, in 2008, and completed his Ph.D. in Biophysics at the University of Michigan, Ann Arbor, USA, in 2013. He received two post-doctoral fellowships from the University of California, Los Angeles (2015–2018) and the National Institute of Standards and Technology (NIST) (2014–2015). He has also worked in many different research labs, including the Max Planck Institute for the science of light in Erlangen, Germany, the University of Arizona, Tucson, and the National Tshinghua University in Taiwan on different applications of photonics. His research interests are utilizing nanotechnology and multi-modal optical imaging for disease diagnostics and treatment. Dr. Ray has received several fellowships including NIH Ruth L. Kirschstein National Research Service Award and the SPIE student scholarship. He has over 30 peer reviewed journal articles and several patents pending.

Hatice Ceylan Koydemir is a senior researcher at the Department of Electrical and Computer Engineering at UCLA. Her research interests include biophotonics, optical sensors, MEMS-based biosensors, micro-fabrication technologies, and lab-on-a-chip devices for point-of-care diagnosis and analysis. Ceylan Koydemir received her B.Sc. degree in Environmental Engineering in 2004, her minor degree in Food Engineering in 2004, M.Sc. and Ph.D. degrees in Chemical Engineering in 2007 and 2013, respectively, from Middle East Technical University (METU) in Turkey. Following her Ph.D., she joined the Ozcan Research Group at UCLA as a postdoctoral researcher. She was a recipient of the METU Ph.D. Thesis of Year Award in 2013, Prof. Dr. Hasan Orbey Ph.D. Thesis Award in 2011 and Dr. Haluk Sanver Technology Award in 2013, which was given by the Department of Chemical Engineering at METU. She has written more than 27 peer-reviewed articles, 75+ conference presentations, a book chapter, as well as one issued and five pending patent applications.

Editorial

Mobile Diagnostic Devices for Digital Transformation in Personalized Healthcare

Hatice Ceylan Koydemir [1],* and Aniruddha Ray [2],*

1   Department of Electrical and Computer Engineering, University of California, 90095 Los Angeles, CA, USA
2   Department of Physics and Astronomy, University of Toledo, 43606 Toledo, OH, USA
*   Correspondence: hceylan@ucla.edu (H.C.K.); Aniruddha.ray@utoledo.edu (A.R.);
    Tel.: +1-(419)530-4787 (A.R.)

Received: 14 November 2020; Accepted: 24 November 2020; Published: 25 November 2020

Mobile devices have increasingly become an essential part of the healthcare system worldwide. This is particularly evident during the current COVID-19 pandemic, as telemedicine is playing an important role in enabling remote interaction between physicians and patients, to support patient care and disease management, while maintaining social distancing. In addition to telemedicine, mobile devices have played a key part in evaluating and guiding the pandemic response, by facilitating contact tracing, as well as helping with mapping the transmission [1]. There has also been a significant effort towards the development of mobile diagnostic devices for point-of-care (POC) testing of COVID-19 in order to curb the pandemic. A few different POC devices for the rapid detection of the SARS-CoV-2 have been developed, including the Accula system (Mesa Biotech), Sofia 2 (Quidel), Talis One (Talis Biomedical) and Cue (Cue Health), which are expected to enable rapid testing and significantly increase the rate of screening. Thus, the use of POC diagnostic devices will play a key role during such pandemics, especially where vaccines are yet to be developed, by preventing the spread of the disease and reducing the mortality rates. Although promising, there are several challenges associated with the deployment of such devices. These include filing regulatory approvals in different countries worldwide, negotiating the terms of cost reimbursement from insurance companies, maintaining data privacy and the protection of data, among others. Despite the challenges, mobile devices offer several advantages that make them attractive to clinicians, patients, and even educators. One of the main advantages of mobile devices is their ability to provide healthcare access to patients in low-resource settings, such as rural areas or poor communities, at an affordable cost. Mobile devices can be used for on-site testing and data collection using their in-built cameras and sensors, as well as external lab-on-chip platforms, paper-based assays, and other formats of POC tests [2]. Advancement in processor technology has also facilitated complicated calculations and analysis, on the spot, using these mobile devices and simple user interfaces (e.g., mobile applications). Furthermore, the data acquired from on-site testing can also be uploaded to different servers worldwide, which would facilitate data analysis using computationally intense approaches, such as machine learning, deep learning, or even crowdsourcing. Another advantage of remote diagnostics and telemedicine is the reduction in patients' exposure to nosocomial infections, which can be life-threatening (e.g., methicillin-resistant *Staphylococcus aureus* (MRSA)). The utilization of mobile devices such as smartphones, smartwatches, implantable sensors, and portable readers in healthcare has several benefits to public health strategies as well, especially for monitoring chronic medical conditions. For example, smartwatches can measure blood pressure in patients suffering from hypertension. Standalone medical devices, such as the glucometer, which is a standard tool for diabetes management, have become quite popular as well. Another important area is the detection of bacterial and viral pathogens, which was reviewed by Nath et al. [3] The pathogen detection devices are typically based on portable lab-on-chip platforms, such as microfluidics or plasmonics, and integrated with optical or electrochemical readers. This review focuses on the point-of-care optical readers, such as smartphone and holographic microscopes, with in vitro diagnostic assays that are easy to perform in resource-limited settings.

Dieng et al. [4] developed a mobile set-up, composed of a sample inactivation extraction station, sample preparation station, and detection station, for the detection of the causative agent of non-malaria febrile illnesses by screening for dengue virus (DENV), Zika virus (ZIKV), yellow fever virus (YFV), chikungunya virus (CHIKV), and rift valley fever virus (RVFV). They evaluated the performance of this mobile laboratory by screening the blood samples of 104 children in Senegal and confirmed their test results to gold standard molecular methods such as real-time polymerase chain reaction (RT-PCR) and enzyme-linked immunosorbent assay (ELISA). Another optical platform for virus detection, i.e., Zika virus, for which effective vaccines are not yet available, was demonstrated by Kabir et al. [5]. This mobile in vitro diagnostic platform is based on a smartphone-based video reader for a colorimetric assay. This device has a high throughput with a 9 min turnaround time, including sample preparation.

Smartphones can also be used in combination with smart wearable sensors and implants to collect patient data, which can then be used to predict different health problems, as well as monitor the dynamics of a particular illness [6]. Rodriguez-Riuz et al. [7] analyzed the night, day, and 24 h motor activity data of 55 patients, of which 23 were diagnosed with depressive episodes, to find the best dataset for the accurate classification of depressive episodes using machine learning and a smart wearable. They demonstrated that the night motor activity data were the best dataset for this classification with >99% specificity and sensitivity. Similarly, smartphones were also used to study neurological disorders and disabilities. Carmono-Perez et al. [8] studied the craniocervical range of motion in subjects between the ages of 4 and 14 with cerebral palsy, which is a neurological disorder that affects movement and motor skills. An inertial measurement unit (IMU) device and a cervical range of motion (CROM) device were used to perform the measurements of movement in different spatial planes. Both IMU and CROM had a high level of correlation between them for the assessment of craniocervical motion. In another example, Shichkina et al. [9] evaluated the effectiveness of the use of the mobile phone and neural networks for the determination of the status of patients with Parkinson's disease. In this study, a smartphone was used to collect patient data, such as speech, hand tremors, tapping of fingers, speed, balance, and reaction time, which was used to train a deep recurrent neural network, that can help determine the condition of the patient. In a separate study, Rodriguez-Almagro [10] used a new virtual reality system, consisting of a mobile device and a headset, to perform subjective visual vertical (SVV) test, on subjects suffering from disabilities such as migraine and tension-type headache. However, this test did not yield any significant differences between the control group and the group with a disability, thereby highlighting the need to perform additional visual and somatosensory tests.

In conclusion, a number of interesting applications ranging from pathogen detection to neurological disorders and mental health problems were presented in this Special Issue, which highlights the capabilities of mobile diagnostic devices and their importance in personalized healthcare.

Funding: This research received no external funding.

Conflicts of Interest: The authors declare no conflict of interest.

References

1. Grantz, K.H.; Meredith, H.R.; Cummings, D.A.T.; Metcalf, C.J.E.; Grenfell, B.T.; Giles, J.R.; Mehta, S.; Solomon, S.; Labrique, A.; Kishore, N.; et al. The use of mobile phone data to inform analysis of COVID-19 pandemic epidemiology. *Nat. Commun.* **2020**, *11*, 4961. [CrossRef] [PubMed]
2. Vashist, S.K.; Luppa, P.B.; Yeo, L.Y.; Ozcan, A.; Luong, J.H.T. Emerging Technologies for Next-Generation Point-of-Care Testing. *Trends Biotechnol.* **2015**, *33*, 692–705. [CrossRef] [PubMed]
3. Nath, P.; Kabir, A.; Doust, S.K.; Kreais, Z.J.; Ray, A. Detection of Bacterial and Viral Pathogens Using Photonic Point-of-Care Devices. *Diagnostics* **2020**, *10*, 841. [CrossRef] [PubMed]
4. Dieng, I.; Hedible, B.G.; Diagne, M.M.; Wahed, A.A.E.; Diagne, C.T.; Fall, C.; Richard, V.; Vray, M.; Weidmann, M.; Faye, O.; et al. Mobile Laboratory Reveals the Circulation of Dengue Virus Serotype I of Asian Origin in Medina Gounass (Guediawaye), Senegal. *Diagnostics* **2020**, *10*, 408. [CrossRef] [PubMed]

5. Kabir, M.A.; Zilouchian, H.; Sher, M.; Asghar, W. Development of a Flow-Free Automated Colorimetric Detection Assay Integrated with Smartphone for Zika NS1. *Diagnostics* **2020**, *10*, 42. [CrossRef] [PubMed]

6. Koydemir, H.C.; Ozcan, A. Wearable and Implantable Sensors for Biomedical Applications. In *Annual Review of Analytical Chemistry*; Bohn, P.W., Pemberton, J.E., Eds.; Annual Reviews: Palo Alto, CA, USA, 2018; Volume 11, pp. 127–146.

7. Rodriguez-Ruiz, J.G.; Galvan-Tejada, C.E.; Zanella-Calzada, L.A.; Celaya-Padilla, J.M.; Galvan-Tejada, J.I.; Gamboa-Rosales, H.; Luna-Garcia, H.; Magallanes-Quintanar, R.; Soto-Murillo, M.A. Comparison of Night, Day and 24 h Motor Activity Data for the Classification of Depressive Episodes. *Diagnostics* **2020**, *10*, 162. [CrossRef] [PubMed]

8. Carmona-Perez, C.; Garrido-Castro, J.L.; Vidal, F.T.; Alcaraz-Clarina, S.; Garcia-Luque, L.; Alburquerque-Sendin, F.; Rodrigues-de-Souza, D.P. Concurrent Validity and Reliability of an Inertial Measurement Unit for the Assessment of Craniocervical Range of Motion in Subjects with Cerebral Palsy. *Diagnostics* **2020**, *10*, 80. [CrossRef] [PubMed]

9. Shichkina, Y.; Stanevich, E.; Irishina, Y. Assessment of the Status of Patients with Parkinson's Disease Using Neural Networks and Mobile Phone Sensors. *Diagnostics* **2020**, *10*, 214. [CrossRef] [PubMed]

10. Rodríguez-Almagro, D.; Obrero-Gaitán, E.; Lomas-Vega, R.; Zagalaz-Anula, N.; Osuna-Pérez, M.C.; Achalandabaso-Ochoa, A. New Mobile Device to Measure Verticality Perception: Results in Young Subjects with Headaches. *Diagnostics* **2020**, *10*, 796. [CrossRef] [PubMed]

Publisher's Note: MDPI stays neutral with regard to jurisdictional claims in published maps and institutional affiliations.

 *diagnostics*

Review

Detection of Bacterial and Viral Pathogens Using Photonic Point-of-Care Devices

Peuli Nath [†], Alamgir Kabir [†], Somaiyeh Khoubafarin Doust, Zachary Joseph Kreais and Aniruddha Ray *

Department of Physics and Astronomy, University of Toledo, Toledo, OH 43606, USA;
Peuli.Nath@UToledo.Edu (P.N.); MdAlamgir.Kabir@rockets.utoledo.edu (A.K.);
Somaiyeh.KhoubafarinDoust@rockets.utoledo.edu (S.K.D.); Zachary.Kreais@rockets.utoledo.edu (Z.J.K.)
* Correspondence: Aniruddha.ray@utoledo.edu; Tel.: +1-(419)-530-4787
† These authors contributed equally.

Received: 7 September 2020; Accepted: 15 October 2020; Published: 19 October 2020

Abstract: Infectious diseases caused by bacteria and viruses are highly contagious and can easily be transmitted via air, water, body fluids, etc. Throughout human civilization, there have been several pandemic outbreaks, such as the Plague, Spanish Flu, Swine-Flu, and, recently, COVID-19, amongst many others. Early diagnosis not only increases the chance of quick recovery but also helps prevent the spread of infections. Conventional diagnostic techniques can provide reliable results but have several drawbacks, including costly devices, lengthy wait time, and requirement of trained professionals to operate the devices, making them inaccessible in low-resource settings. Thus, a significant effort has been directed towards point-of-care (POC) devices that enable rapid diagnosis of bacterial and viral infections. A majority of the POC devices are based on plasmonics and/or microfluidics-based platforms integrated with mobile readers and imaging systems. These techniques have been shown to provide rapid, sensitive detection of pathogens. The advantages of POC devices include low-cost, rapid results, and portability, which enables on-site testing anywhere across the globe. Here we aim to review the recent advances in novel POC technologies in detecting bacteria and viruses that led to a breakthrough in the modern healthcare industry.

Keywords: infectious diseases; diagnostics; point-of-care devices; microfluidics; plasmonics; smartphone; lensless imaging

---

1. Introduction

Throughout human history, there have been several epidemics and pandemics, worldwide, due to the emergence and re-emergence of disease causing microorganisms, such as bacteria and viruses [1–4]. These pathogens cause diseases that are contagious and can be transmitted easily via aerosols, food, physical contact, body fluids of the infected person in a short period of time [5]. Infectious diseases like HIV (Human immunodeficiency virus), SARS (Severe acute respiratory syndrome), COVID-19 (coronavirus disease 2019 caused by SARS-CoV-2), influenza flu, Ebola, Herpes, Hepatitis, and tuberculosis are some of the top global health challenges at present [6,7]. For example, acquired immunodeficiency syndrome (AIDS) caused by HIV, had affected nearly 37 million people (including 1.8 million new infections) across the globe by the end of 2017 [8]. Currently, there is no cure for AIDS; early detection and prevention of transmission is the only way to defeat this disease. Some of the other pandemics include the 2009 swine flu, caused by the H1N1 Influenza virus that affected approximately 61 million people in the US alone; the 1918 Spanish flu, which resulted in the loss of almost 50 million lives worldwide [9,10]; and, more recently, COVID-19, which has infected more than 25 million people to date, with nearly ~0.8 million deaths worldwide [11,12]. The rapid spread of this infection across the continents brought life to a complete standstill. Additionally, these epidemics and pandemics have a severe impact on the economy. The COVID-19 pandemic also resulted in a severe decline of the world economy, including a 5% reduction in the GDP of the US [13]. Among bacterial infections, tuberculosis (TB) is one of the deadliest diseases caused by a bacteria *Mycobacterium tuberculosis*, which has resulted in over 1.2 million deaths in 2018 alone. It is more prevalent in countries like India, Nigeria, Indonesia, and Philippines, which account for half of the ~10 million global cases [14]. According to WHO, the estimated treatment coverage rate in 2018 was 69%, and the major challenge lies in its rapid diagnosis [14–18]. Foodborne pathogens, such as some virulent strains of *Escherichia coli*, are known to cause many diseases like colitis, urinary tract infections, meningitis, sepsis, and many more [19–21]. Every year, more than five million deaths occur worldwide due to these diseases especially in low- and middle-income countries [22]. The elderly population and children with underdeveloped immunity are particularly vulnerable to these infections.

Thus, there has been a constant focus on early detection of these pathogens with high sensitivity and specificity, in order to prevent the spread of these infections. Some of the commonly used techniques for detection of bacteria and viruses include blood culture, high-throughput immunoassays, e.g., enzyme-linked immunosorbent assay (ELISA), polymerase chain reaction (PCR), mass spectrometry (MS), etc. [23–25]. Although the conventional diagnostic methods provide accurate results, they often lack sensitivity, are-time consuming, expensive, require intensive labor for sample preparation, and need trained laboratory personnel to carry out the tests. Availability of these diagnostic tests to the general population is a major issue, which has been highlighted during the COVID-19 pandemic. At present due to lack of resources, there is a strong preference to mainly test people with COVID-19 symptoms, thereby bypassing a majority of the asymptomatic population [26]. Additionally, it takes several days to get the results back from the clinic. This problem is particularly severe in developing countries, where patients usually need to travel to diagnostic centers and wait long hours in hospitals to get the test done. It has been previously reported that in developing nations over 95% deaths occur due to lack of proper diagnosis and treatment [27].

In order to address the aforementioned shortcomings, a herculean effort has been directed towards developing point-of-care (POC) devices. These devices are able to provide diagnosis at the point of care, without the need to go to a clinical laboratory [28–30]. An ideal POC device, such as the glucometer, can be used for testing patients at the comfort of their home, with minimal or no supervision, and should be able to provide results rapidly. These devices are designed to be low in cost, portable, and easy to use [28]. A simple POC device relies on a (i) biological recognition element (enzyme, proteins, antibody, and aptamer) that selectively interacts with the target molecules (antigen), and (ii) a transducer that monitors the interaction and provide information both qualitatively and quantitatively. Typically, the POC devices or biosensors are developed by integrating plasmonic or microfluidic devices, and an electrochemical or optical readout system into a single miniaturized platform for real-time detection of pathogens, as shown in Figure 1 [31–34]. In resource-limited settings, a POC device promises (ASSURED) Affordable, Specific, Sensitive, User-friendly, Rapid, Equipment-free analysis of immunoassays (antigen and antibody reaction) and Delivery to remote areas for 'on-site' analysis of samples, according to the guidelines set by WHO for developing diagnostic tools for economically underdeveloped nations, to enhance global healthcare quality [35].

In this review, we focus on some of the commonly used technologies utilized for developing POC devices for the detection of bacteria and viruses. These include microfluidics, plasmonics, smartphone-based imagers and lensless microscopes (Figure 1). We focus mainly on photonics-based technologies as they are capable of extremely sensitive measurement at a very high resolution and the ability to operate in multiple different modalities, e.g., colorimetric, transmission, scattering, reflection, fluorescence, interferometry, etc. Diagnosis of the diseases involves either direct detection of the pathogen or indirect detection of the antibodies produced in the body in response to a particular microorganism. We discuss some of the specific examples in detail and highlight the current state of various POC devices developed over the past decade.

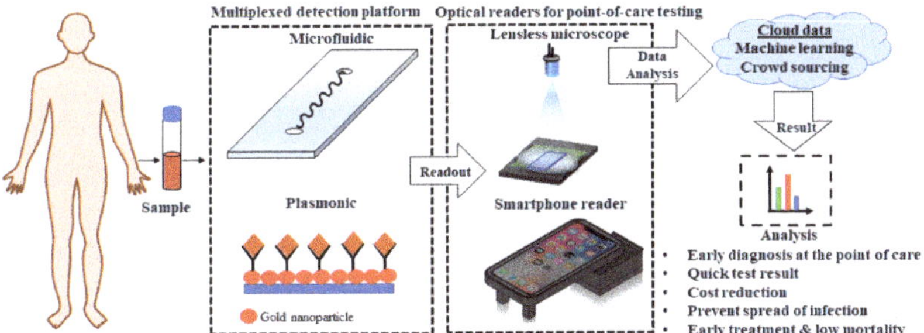

Figure 1. Flow diagram showing the process of sample testing using point-of care devices.

2. Microfluidics-Based Platforms

Ever since the development of the first commercial devices μ-TAS in 1990, microfluidic technologies have evolved significantly and has been used for a large number of medical diagnostic applications [34]. Microfluidics is a field of research that deals with the manipulation of fluids at the microscale inside channels of dimension less than 1000 micron [36]. It provides the advantage of setting up experiments that require rapid diffusion, laminar flow, small sample volume, and large surface-area-to-volume ratio. The miniature size of the devices and the requirement of small sample volume makes it ideal for point of care applications. Additionally, these platforms can be used to support many different assays including immunoassays, nucleic acid amplification assays, and biochemical reactions. Therefore, microfluidics is frequently incorporated in point-of-care diagnostic devices [37,38].

A microfluidic (MF) platform is usually developed by using materials that are lightweight, inexpensive, portable, and disposable, such as polymers, glass, paper, and textiles, among others. Each of the materials has its own unique advantages [39–42]. For example, paper microfluidics is one of the most extensively used platforms that has been used for a variety of bio-analyte detection, due to its easy availability, low cost, biodegradability, portability, lightweight nature, and self-capillary action that eliminates the need for an external pump. The paper-based MF device is a common candidate for lateral flow immunoassays (LFIA)/test strip/rapid test/dipstick device for the detection of pathogens, antigens, and antibodies [43–45]. LFIAs generally consists of a sample loading pad, absorbent pad, conjugate pad, a test line, and a control line on a membrane (commonly used nitrocellulose membrane). One example of the LFIA is the recently developed tuberculosis detection, where the sample is deposited on the loading pad and flows laterally to meet the conjugate pad that contains immobilized gold nanoparticles (AuNPs) tagged with antibody (Ab) that specifically captures the CFP10-ESAT6 antigen of *M. tuberculosis* in the sample [46]. The AuNP-Ab-Antigen complex then flows along the membrane laterally due to the self-capillary action of the membrane and meets the test line, which has a second antibody that captures the AuNP-Ab-Antigen complex, resulting in a colored line. Vertical flow immunoassays (VFIA) are alternate paper-based assays that are based on the vertical flow of sample due to gravity and capillary action, and they tend to have a faster detection time [45]. In addition to papers, glass-based MF devices are also quite common due to their excellent optical properties, chemical inertness, surface stability, and solvent compatibility; thus, glass is used for fabricating devices for the detection of enzymes, antibodies, and whole cell [47]. Polymer-based MF devices fabricated by using polydimethylsiloxane (PDMS), polyethylene, polypropylene, etc., are extensively used in commercial devices due to their low cost, compared to glass and silicon, high transparency, and chemical/electrical resistance which is particularly desirable for electrochemical immunosensors [42]. Other polymers such as thermoplastics (polystyrene, cyclin olefin copolymer (COC), polyethylene terephthalate (PET), poly (methyl methacrylate) (PMMA), and polycarbonate) are also used for large scale microfluidic chip fabrication. These polymers are rigid, have good mechanical strength with no deformation issues, low water absorption, high chemical resistivity, and excellent optical properties with high UV transparency [48]. Microfluidic devices can be fabricated by using several different techniques, depending on the material. Photolithography is the common method for fabricating microfluidic devices; other methods include micromachining, plasma etching, hot embossing, injection molding, 3D printing, laser ablation, and, recently, nanofabrication [47,49]. The type of fabrication depends on the material, as well as the specific application. PDMS-based microfluidic devices are fabricated by using a soft lithography technique where liquid PDMS is poured in a micro-mold (SU-8), followed by curing at high temperature (60–80 °C) for 2 h. Meanwhile, thermoplastic polymer-based chips are fabricated in two ways: rapid prototyping and replication methods. In rapid prototyping, computer numerical controlled (CNC) machine and laser ablation techniques are employed. For large-scale production of microchannels with thermoplastic substrate, replication methods such as hot embossing, imprinting, and injection molding are commonly used. Unlike PDMS, the bonding step in thermoplastic microfluidic devices is critical. Typically, direct bonding includes thermal fusion bonding, ultrasonic bonding, and surface modification, whereas indirect bonding involves the use of chemical reagents, such as epoxy, and adhesive tape, to assist the bonding. Recent advancement in microfluidic technology includes the development of 'hybrid devices', i.e., integrating PDMS or paper with thermoplastic such as PDMS–PET/PDMS–PMMA [49].

Microfluidic platforms have been used for the detection of a variety of different pathogens that causes some of the deadliest bacterial and viral diseases such as influenza, human immunodeficiency virus (HIV), tuberculosis, Hepatitis B, Ebola, Hepatitis C, and food poisoning [50–57]. The use of microfluidic technologies is heavily featured in the newly developed POC devices for COVID-19. These include the Accula system (Mesa Biotech), which utilized the RT-PCR process; Talis One (Talis Biomedical), which is based on the loop-mediated isothermal amplification (LAMP) technology; and the Sofia 2 (Quidel), which is based on the detection of viral proteins using fluorescence [58]. Another lateral flow test for POC detection of SARS-CoV-2 was developed by combining isothermal amplification and CRISPR mediated detection method (SHERLOCK: Specific High-Sensitivity Enzymatic Reporter UnLocking) [59]. The SHERLOCK technology involves the detection of DNA or RNA by amplification of viral genome by an isothermal amplification assay and detection of the amplicon by CRISPR mediated reporter unlocking. This test called STOP (SHERLOCK Testing in One Pot) was developed to eliminate the need for sample extraction and complex reagent handling, and it can be operated at a single temperature. The best LAMP primers are designed for optimal amplification targeting N (nucleoprotein) gene. Cas12b enzyme from *Alicyclobacillus acidiphilus* (AapCas12b) was explored and operated at an optimum temperature of 55 °C for the one-pot reaction. As AapCas12b did not contain CRISPR array; 18sgRNA AacCas12b which has 97% identical sequence was combined with AapCas12b. The one-pot reaction generated faster results with higher collateral activity. The test results were generated within one hour and comparable to the standard RT-PCR technique with a limit of detection (LOD) of 100 copies of the viral genome and have been validated, using COVID-19 patient samples.

A different microfluidic platform was developed for the detection of the influenza A (H1N1) virus by using an electrochemical approach, involving an electrochemical immunosensor coated with reduced graphene oxide (RGO) [60]. A PDMS microfluidic channel was fabricated with a thickness of 200 μm and height of 100 μm and has three electrode settings with Au-WE (gold working electrode), where the immunobinding takes place; Pt-RE (platinum reference electrode) as a stable potential reference; and Au-CE (gold counter electrode), which collects the current between WE and itself. Glass coverslips were used as a support for the three electrodes. The electrodes were coated with RGO, using dip-coating method, and, subsequently, a monoclonal Ab (mAb) specific to H1N1 virus was attached to the carboxyl group (COOH) of RGO via EDC/NHS (1-ethyl-3-(-3-dimethylaminopropyl) carbodiimide/N-hydroxysuccinimide) coupling. The binding of the H1N1 virus with the mAb attached on the electrode resulted in a voltage change, which was monitored by using cyclic voltammetry. The limit of detection (LOD) of this approach was 0.5 PFU/mL, with a linear concentration range of $1–10^4$ plaque forming unit/mL (PFU/mL), which is better than most other immunosensors developed so far.

The detection of the Zika virus (ZIKV), which became a major global health concern in the year 2015/2016, was another challenge that was addressed using a smartphone-based fluorescent lateral flow immunoassay POC device for the detection of the non-structural protein (NS1) of ZIKV [61]. Fluorescent quantum dots (QDs) conjugated with the ZIKV NS1 antibody were used as the detection antibody in the absorption pad, mouse monoclonal ZIKV NS1 antibody in the test line as the capture antibody, and polyclonal goat anti-mouse IgG antibody in the control line of the LFIA. In presence of NS1 antigen, fluorescent QDs-ZIKV NS1 antibody captured the antigen and then flowed laterally along the nitrocellulose membrane and form QD-Ab-NS1-Ab sandwich complex on the test line. The luorescence signal was recorded using a smartphone, under a hand-held UV lamp at 365 nm, and analyzed for quantitative detection of ZIKV NS1 antigen. The assay could detect up to 0.15 ng/mL NS1 in serum in under 20 min. Another type of automated POC microfluidic device was developed based on the colorimetric detection of ZIKV NS1 protein using ELISA assay as shown in Figure 2 [62]. A 3-layer disposable POC microfluidic chip was fabricated using polymethylmethacrylate (PMMA) and double-sided adhesive tape. The 750 μm thick top layer have inlets and outlets for sample loading. The 1.5 mm thick middle layer contained all the reagents and aqueous solution followed by the solid bottom layer which acted as the base support for the microfluidic chip. The microfluidic chip was loaded with all the reagents (Phosphate buffer, washing buffer, blocking buffer, and 3,3′,5,5′ tetramethylene blue (TMB) solution) in different chambers. The magnetic microfluidic ELISA (M-ELISA) assay involved the use of magnetic particles which were coated with biotinylated ZIKV NS1 capture antibody via neutravidin, present on the particles. The ZIKV NS1 antigen was captured using the antibody conjugated magnetic beads and transferred onto the chip. The chip was placed in a magnetic actuator platform to automatically perform washing, binding to horseradish peroxidase (HRP) tagged anti-ZIKV NS1 antibody, which completed the sandwich structure. TMB was used to generate a blue colored product and quantify the viral concentration. The magnetic actuator platform consisted of an Arduino controlling unit and a 3D printed platform that accommodated the microfluidic chip as shown in Figure 2. An iPhone X was used to capture the video, which was used for analysis based on color intensity. The limit of detection, when using this M-ELISA on-chip technique, was found to be 62.5 ng/mL in whole plasma, which is better than any other reported ELISA-based POC devices for ZIKV NS1 detection.

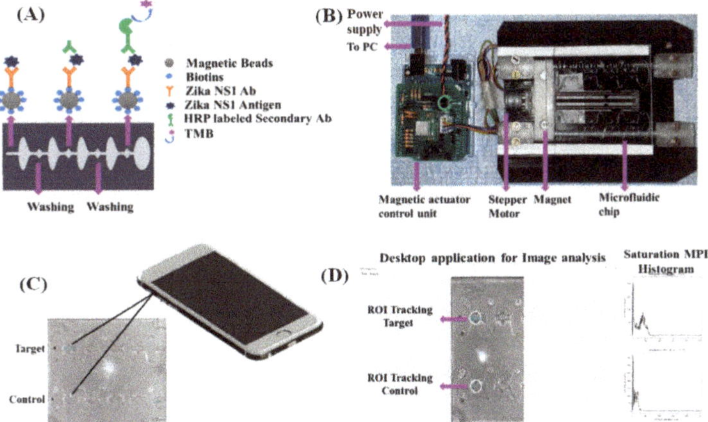

Figure 2. (**A**) Schematic of the enzyme-linked immunosorbent assay (M-ELISA) inside the microfluidic chip; (**B**) Magnetic actuation platform holding the microfluidic chip controlled by Arduino controller allowing bi-directional movement of the magnets; (**C**) Colorimetric changes in the chip were recorded using a smartphone; (**D**) Histogram plot of the saturation maximum pixel intensity (MPI) of the color following the M-ELISA assay on chip [62].

Recently, loop-mediated isothermal amplification technique has been used for simple, rapid and accurate detection of positive sense single-stranded RNA virus ZIKV [63,64]. Kaarj et al. demonstrated the RT-LAMP technique on a simple disposable platform that included 'paper microfluidics' coupled with pH-indicator-based colorimetric assays integrated with a smartphone reader [55]. Cellulose paper, owing to its negative polarity, can be used for separating target RNA fragments from other proteins present in blood plasma of infected samples, based on their size and charge, thereby minimizing the need for pretreatment of the samples. The paper microfluidic chip was developed by using various types of material with different pore sizes, e.g., nitrocellulose (NC) paper, and grade 4 (G4) and grade 1 (G1) cellulose paper. The sample loading area (5 × 5 mm) was connected to the main channel (3 × 30 mm), where filtration occurred, and was followed by the detection zone (5 mm diameter), where the target RNA fragments were collected. After collecting the target RNA fragments the detection zone was cut out, loaded with RT-LAMP reagent mixture, sandwiched between two glass slides, and sealed with parafilm, to prevent sample evaporation. The RT-LAMP mixer had the primer and colorimetric dye, a pH indicator phenol red. The primer was designed to bind specifically to only the NS5 gene of ZIKV. The detection zone was then placed on a hot plate (68 °C) for 30 min, resulting in the amplification of the RNA, which led to a change in color from yellowish red to yellow (in the presence of the ZIKV). A smartphone-based reader was used to monitor the color change and analyze the images by using the ratio of red to green pixels. This RT-LAMP assay, using a paper microfluidic chip, can detect ZIKV with limit of detection (LOD) as low as 1 copy/µL.

HIV, which is one of the most severe global healthcare challenges over the past few decades, has attracted a lot of attention, and several commercially available rapid diagnostic tests have been developed. One of the first commercially available FDA-approved rapid diagnostics test for HIV was the Murex single-use diagnostic system (Abbott Laboratories). However, the test generated too many false results when compared to the conventional ELISA technique [65]. A similar study was conducted using five commercially available fourth and fifth generation ELISA kits for HIV detection, using different batches of confirmed number of positive and negative samples (100 in total) to evaluate the testing quality. None of the evaluated ELISA kits were able to identify all the samples correctly with 100% efficiency across all the batches, but showed high sensitivity; however, they have low specificity, particularly in the initial phases of the infection [66]. There are portable LFIA-based POC devices available in the market, such as Ora Quick Rapid in home HIV1/2 Antibody test, which can detect HIV antigen, even at low concentrations, from oral fluid, and suitable is for on-site analysis. Despite having high specificity, the test has a sensitivity of only ~92%, thereby generating few false-negative results [67]. The gold standard method for detecting HIV antigen is RT-PCR, but it can only be performed in a laboratory [68]. Recently, Phillips et al. developed a fully microfluidic rapid and autonomous analysis device (micro-RAAD) for the detection of HIV from whole-blood sample, based on loop-mediated isothermal amplification (LAMP) of HIV RNA [69]. The microfluidic device consisted of two different paper membranes: The first was the blood-separation membrane, and the second was an amplification membrane that isolated the HIV viral proteins present in blood. RNA from the isolated viral particle was amplified by using RT-LAMP reagents coated on the paper membrane which target the *gag* gene of the HIV-1 and presented the amplicons to the attached LFIA for visualization. The device was connected to a reusable temperature circuit and could be operated using a laptop or smartphone. The sensitivity of this integrated prototype was 3×10^5 virus copies per reaction, or 2.3×10^7 virus copies per mL of whole blood, which is comparable to clinically reported HIV-1 concentration at the peak of infection [70].

A different microfluidic diagnostic assay platform containing multiple detection modalities was developed by Shafiee et al., for the detection of different bio-analytes (both viruses and bacteria) from whole blood, serum, and other bodily fluids, with high specificity and sensitivity [71]. For HIV-1 detection, a microfluidic channel was fabricated by using a flexible substrate polyester film with two silver electrodes, using silver ink for detection of the virus, using viral lysate impedance spectroscopy, as shown in Figure 3A. The platform has three layers: top and bottom transparent substrate (polyester) layers and double-sided adhesive (DSA) film between the channel layer. The inlets and outlets were cut on the polyester film, with a diameter of 0.6 mm, and channels were cut on the DSA, using a laser cutter. The ink was poured through the polyester inlets, to fill the openings evenly on the DSA, using a glass coverslip. After the ink dried, the DSA was removed, leaving the electrodes on the polyester film substrate. The dimension of the electrodes was 2 mm × 1 mm. For the assay, polyclonal anti-gp120 antibody-conjugated magnetic beads were used. The HIV-1 virus was first isolated and captured by using the Ab-coated magnetic beads and detected by using the impedance magnitude measurement of the viral lysate samples. Viral lysis increased the electrical conductivity and decreases the bulk impedance magnitude of the solution. Impedance magnitude and signals were measured at 1 V with pre-evaluated frequencies between 100 Hz and 1 MHz. An electronic circuit was developed that generated an electrical response of the viral lysate in the microfluidic channel. The viral load of different subtypes was predetermined. The test samples were prepared by spiking whole blood with a cocktail of HIV-1 subtypes (A, B, C, D, E, and G). The control samples were prepared with HIV-free phosphate buffer and magnetic beads. The system could effectively detect HIV-1 virus at concentrations upwards of ~10^6 copies/mL.

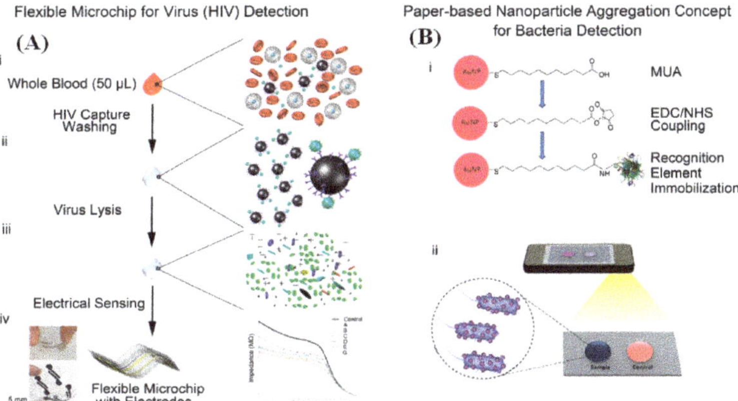

Figure 3. (**A**) Schematic of the flexible polyester film-based electrical sensing platform for HIV detection, including the capture of HIV through the use of anti-gp120 antibody coated magnetic beads, washing and lysis steps, and measurement of electrical impedance; (**B**) Detection of bacteria on cellulose paper, using a smartphone, based on nanoparticle aggregation assay. The following schematics depict the gold nanoparticle surface modification steps and the resultant aggregation assay which is detected by using a smartphone [71].

In addition to the HIV-1, they also fabricated a paper-based nanoparticle aggregation assay system incorporated with a smartphone reader platform for detection of *E. coli* in whole blood, serum, and other bodily fluids with high specificity and sensitivity [71].

The device for *E. coli* detection involves the use of cellulose paper modified with nanoparticle and subsequent imaging with a smartphone as shown in Figure 3B. For *E. coli* detection, gold nanoparticle (AuNP) was modified with 11-mercaptoundecanoic acid (MUA), and succinimide groups were generated by using EDC/NHS mixture for attachment of the amine-terminated proteins on the MUA-AuNP surface. For effective binding of *E. coli* to AuNP, liposaccharide binding protein (LBP) was added to the AuNP-MUA solution. The *E. coli* spiked samples along with AuNP solution was then added to the cellulose paper, using the drop method and dried. The dried paper was then placed in a black box and illuminated with LED light. This test was based on the nanoparticle aggregation assay. The presence of *E. coli* resulted in the aggregation of the AuNPs, causing a visual color change of AuNP from red to blue. Images were captured, using a smartphone, and the individual RGB values were used for analysis. The limit of detection of this assay was reported to be 8 colony forming unit/mL (CFU/mL).

Tuberculosis (TB) is another deadly bacterial disease that has attracted a lot of attention. Commercially available ELISA systems for the detection of tuberculosis are based on the interferon gamma release assay (IGRA). The T cells in TB infected patients produces a pro-inflammatory cytokine, interferon-gamma (IFNγ) in response to TB specific antigens, which is used for diagnosis purpose. These IFNγ producing T cells are quantified by using ELISA spot [72,73]. However, the process (IGRA) is time-consuming and requires pre-incubation of blood with TB antigens before sample preparation. Modifying this technology, Evans et al. fabricated a low-cost ELISA amperometric detection unit, using commercially available lab-on-a-chip printed circuit board (PCB) integrated with a microfluidic channel and electrodes attached to the PCB surface for the detection of cytokine IFNγ, with high sensitivity [74]. The device consists of a gold (Au) electrode sensor chip surface in the microfluidic channel that was immobilized with capture antibody anti-IFNγ Fab′(Cys)3. Samples with IFNγ was then flowed (flow rate 25 μL/min, 4 min) across the channel over the sensor chip and plasmon resonance unit (RU) spectra were recorded. The assay involves the use of 3,3′,5,5′ tetra-methylene blue (TMB) as the chromogenic substrate as it is both electrochemically and optically active molecule. TMB acts as a hydrogen donor by enzymatic reduction of hydrogen peroxide by the horseradish-peroxidase-enzyme (HRP) and hydrogen-peroxide-producing free hydroxyl ions and a blue-colored product. TMB is a colorless substrate, but the resulting product (di-imine) is bright blue in color and as the pH is lowered, a change in color from blue to yellow was observed. The absorbance, measured at 450 nm by using a nanodrop spectrophotometer, was used to quantify the change. For the electrochemical assay, fluid wells (50 μL volume) were fabricated by using polymethyl methacrylate (PMMA) over the two Au electrodes fixed to the PCB surface, and reference electrode Ag/AgCl was introduced. The change in current flow due to the electrochemical reactions at the electrode surface was measured by the in-house electronic unit. This assay was able to detect IFNγ with just 30 μL samples at concentrations ranging from 10 to 2000 pg/mL.

All the aforementioned studies demonstrated the capability of microfluidics as a powerful tool to design affordable and disposable POC devices for the detection of a broad range of bacterial and viral pathogens, with high sensitivity and specificity.

3. Surface-Plasmon-Resonance-Based Platforms

Surface plasmon resonance is another diagnostic platform that has been extensively used for the detection of viruses and bacteria [75,76]. Surface plasmons are oscillating electrons on the metallic surfaces that can be excited by shining specific wavelengths of light under certain configurations. Typically, surface plasmon resonance (SPR) is achieved by exciting electrons using evanescent waves, via total internal reflection using prisms. The propagation of evanescent waves is highly dependent on the refractive index of the material (dielectric medium) surrounding the metal surface. SPR biosensors [77,78] typically employ a thin metal film usually gold, silver or aluminum, where biorecognition elements (e.g., antibody, aptamers, etc.) are attached, and plasmons are excited on its surface by the light wave. The binding of the pathogens to the recognition element on the metallic surface leads to an increase in the refractive index of the medium, thereby changing the propagation constant of the surface plasmons. This change of refractive index is measured either by monitoring the change in the resonance angle or the shift in excitation wavelength (Figure 4). This also enables the study of binding affinity and kinetics in real time. Materials with negative real permittivity, such as gold and silver, support surface plasmon polariton (photon strongly coupled to an electric dipole) and thus show the plasmonic activity. The most commonly used material for SPR-based biosensors is gold, as it can be easily functionalized with thiol (-SH) group for surface modification and immobilize of antibodies [78]. The materials used can be of different shapes and sizes, e.g., nanoparticles (spheres, rods, and pyramids), thin films, etc. For localized surface plasmon resonance (LSPR) (Figure 4), nanoparticles smaller than the wavelength of light is typically used [79]. The use of nanoparticles effectively localizes the surface plasmons, and the evanescent wave can extend up to a few tens of nanometer into the sensing medium. In planner surface plasmon resonance (PSPR), a thin film of metal (sheet) is used instead of nanoparticles [33]. The damping of the evanescent wave is less; thus, the penetration depth becomes quite large. Therefore, a living organism can also be studied by using PSPR with a high figure of merit (FOM). However, compared to SPR, LSPR is more sensitive near the surface because of the localized field. These SPR-based techniques can be used for label-free real-time detection of analyte without external labeling (e.g., fluorescent dye, enzymes, etc.) and thus hold extreme potential in POC biosensing [80]. Another approach of utilizing SPR is by exploiting the surface-enhanced Raman scattering (SERS) signal from the target analytes. In SERS-based sensors, the analytes are adsorbed on to corrugated metallic surfaces which results in several orders of magnitude (~routinely 10^6) enhancement of the Raman signatures from the analytes [81,82]. The enhancement of SERS signal results from the strong localization and amplification of the electromagnetic field at the hot-spots on the metallic surface [83–87]. These SERS-based platforms provide an alternate label-free modality for fingerprinting a range of analytes [87].

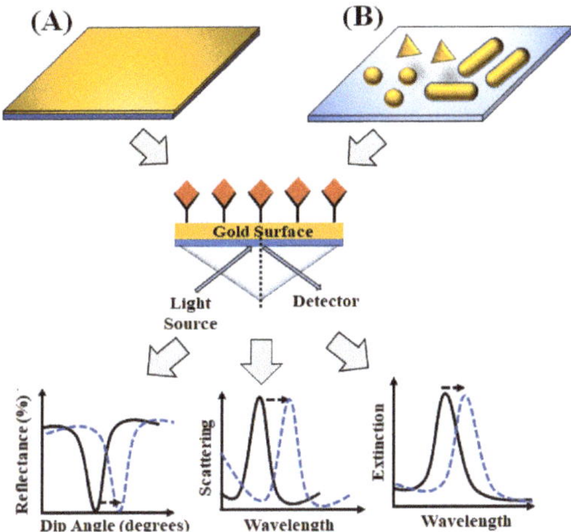

Figure 4. Representation of plasmon-based sensors and the different detection methods: (**A**) planar metallic thin-film-based biosensors and (**B**) localized surface plasmon resonance (LSPR)-based biosensors.

The SPR-based sensors are either stand-alone on-chip platforms, with glass or flexible polymer substrates coated with metallic films/nanoparticle, or they can be integrated with microfluidic platforms containing nanoparticles. Each has its own advantages. Combining the SPR-based sensors with microfluidics facilitates automated testing in small sample volumes [88–90]. Among different types of SPR-based microfluidic devices, flow through SPR sensors are quite popular, particularly in proteomics and drug discovery [90]. In this platform, an SPR sensor is integrated with continuous flow-through channels that enable the detection of multiple analytes as they flow through.

Most of the commercialized flow cell systems use a single inlet and outlet, thus only one sample can be tested at a given time. However, this can be easily addressed by using multiple microchambers for testing of several different analytes in parallel. However, the requirement of multiple valves to prevent cross flow makes it a bit complex. This platform can also be miniaturized by using portable waveguide based SPR sensors where the microfluidic channels are etched on top of the waveguide cladding. LSPR-based biosensing can also be easily performed using lateral flow test strips (discussed in Section 2), by coating nanostructures decorated with target antibodies on self-capillary flow sensor materials, like paper or membrane [33,85,86,91–94]. Digital microfluidics using electrowetting on dielectric (EWOD) is an alternate for continuous flow system where the surface property is controlled by applying a voltage and the contact angle of the droplet can be easily manipulated on the SPR sensor.

This method can be easily employed for automated testing, which includes dispensing, mixing, and separating with enhanced sensitivity [90].

Another promising SPR biosensor for POC applications is based on optical fibers made of silica or polymers [89,91,94–96]. Light propagates inside optical fibers via total internal reflection and the evanescent field on the surface is used to excite the plasmons on the outer metallic coating. Both multi-mode and single-mode fibers can be used, but the latter has a higher sensitivity [97]. Antibodies or aptamers specific to the target antigen is conjugated to the noble metal coated on the fiber surface. The advantage of using optical fibers are manifold. Firstly, due to their flexible nature, they can be used for remote sensing applications and can be designed to operate on a small sample volume. Secondly, utilizing optical fibers reduces the complexity of the devices, by eliminating conventional optical components, thus facilitating miniaturization of the biosensors and improving its portability [97].

Plasmonic platforms have been extensively used for the detection of many pathogens [89,98–105]. The first clinically relevant nano-plasmonic POC platform for the detection and quantification of intact human immunodeficiency viruses (HIV) from unprocessed whole blood cell, with high sensitivity and specificity, was fabricated by Inci et al. [106]. The sensor was prepared by using gold nanoparticles coupled to the anti-gp120 antibody for binding the HIV. Prior to antibody conjugation, the gold nanoparticles were adsorbed onto a polystyrene surface coated with poly-L-lysine. The gold nanoparticles were coated with NeutrAvidin and the biotinylated antibody was conjugated to the particles via the biotin–avidin bond. The presence of the virus in patient blood was detected based on the shift in LSPR wavelength. This sensor can capture and quantify ~98 ± 39 copies/mL in around 1 h and could detect several subtypes of HIV in unprocessed whole blood, making it ideal for POC application.

A bio-plasmonic paper-based device (BPD) was developed for the detection of the ZIKA virus, by quantifying the amount of anti-ZIKV-nonstructural protein 1(NS1) IgG and IgM antibodies in serum [107]. The ZIKV-NS1 protein-coated gold nanorod (AuNR of length ~63 nm and diameter ~25 nm) was used to capture the antibodies. Gold nanorod was used as transducer due to its high refractive index tunability and sensitivity. ZIKV-NS1 was conjugated to the gold nanorod, using a carbodiimide crosslinker and thiol-terminated bifunctional polyethylene glycol (SH-PEG). The BPD was prepared by soaking laboratory filter paper in ZIKV-NS1 functionalized AuNRs solution. The SPR wavelength shift for the ZIKA negative serum (control, $n = 5$) was observed to be 2 nm due to nonspecific binding, whereas a shift of 7.3–8.0 nm was observed for ZIKA positive serum samples ($n = 4$). The metal–organic framework (MOF)-based preservation method rendered the device stable for a month, even at room temperature.

A different type of intensity-modulated surface plasmon resonance (IM-SPR) biosensor was developed by Chang et al. for rapid detection of avian influenza A H7N9 virus [104]. A reaction spot containing the antibody H7-mAb was used to capture the virus. The antibodies were bound to the substrate via amine coupling with the self-assembled monolayer of 11-mercaptoundecanoic acid (MUA) and 6-Mercapto-1-hexanol (MCH) (molar ratio MUA: MCH = 1:9). A reference signal from a second spot was used to quantify the test signal. A polarized light source (laser), at 635 nm, was used to illuminate the two spots and the reflected light was measured, using a data-acquisition device (DAQ). The binding of the antibody-antigen was quantified by using the change in intensity of the reflected light. This simple system was able to detect ~144 viral copies/mL in less than 10 min.

A different approach of measuring the change in intensity due to antibody-antigen binding was demonstrated for the detection of bacteria *E. coli* [108]. In this approach the binding of the bacteria to the *E. coli* O157:H7 antibodies on the gold surface was detected by monitoring the change in photoelectric signal, associated with the SPR shift, using a linear charged coupled devices (CCD). Based on the calibration curve prepared using known quantities of bacteria, the theoretical detection limit was calculated to be 1.87×10^3 CFU/mL. This sensitivity is ~4 times greater than the standard ELISA assays used to detect *E. coli* bacteria.

The change of SPR angle due to antigen–antibody binding is another approach which has been used for the detection of tuberculosis [109]. A portable SPR device was fabricated by Trzaskowski et al. by surface modification of a miniature SPR sensor Spreeta 2000 (S2k) chip with *Mycobacterium tuberculosis* (MTB) antibodies (MPT64 anti-Ag85). The SPR angle was recorded before and after adding the sample. The change in the SPR angle after the binding of MTB was used to quantify the concentration of the bacteria. The detection limit was found to be 1×10^4 CFU/mL for cultured cells but in the sputum sample, it could detect secretory protein at concentration down of ~10 ng/mL. Other approaches, such as the detection of DNA fragment IS6110 using SPR, have also been used for the detection of MTB.

Recently, an SPR-based biosensor was used to detect the SARS-CoV-2 virus. Qiu et al. developed a dual functional plasmonic biosensor combining plasmonic photothermal (PPT) effect and LSPR. The system consisted of two-dimensional gold nano-islands (AuNIs) functionalized with complementary DNA receptors for the detection of the selected sequence of the SARS-CoV-2 virus using nucleic acid hybridization technique. When the system is illuminated, a localized thermo-plasmonic heat is generated, further facilitating the nucleic acid hybridization process and enhancing the selectivity of the assay. The limit of detection of this LSPR-based detection platform was found to be ~0.22 pM [110]. Although this is a benchtop system, this technology can be translated for point-of-care application as well.

Thus, the SPR technique is a very sensitive technique that can be used to detect any bacteria or virus, with very high sensitivity, and can be integrated into any miniaturized POC device.

4. Smartphone-Based Detection System

Modern smartphones with high-quality cameras and excellent computational power have the ability to perform complex analyses, making them ideal candidates for use in point-of-care (POC) devices [111]. Their ease of use and growing popularity in the modern world give smartphone credence to be used anywhere worldwide. In fact, the emerging field of smartphone-based clinical diagnostic devices has the potential to decentralize laboratory and clinical testing, as it offers practical features, such as cost-effectiveness, portability, and building connectivity between patients and healthcare providers. Modern-day smartphones are capable of detecting minute changes in optical signal resulting from any assay including immunoassays, colorimetric assays, and nucleic acid amplification.

Advancements in different fields, such as molecular analysis, biosensors, mathematical algorithms, microfabrication, 3D-printing, and microfluidics, which occur simultaneously with the progress of cellphones and cameras, make it possible to convert a smartphone to a portable diagnostic device. Most smartphone-based diagnostic devices are designed to reduce costs and increase portability, e.g., smartphone-based microscopes and readers. The addition of an external lens, such as ball lenses, to the smartphone camera, can convert the smartphone to a brightfield microscope [112,113]. However, the curved nature of the ball-type lens can cause a distortion around the edge of the image, which can be corrected by using an objective lens and eyepiece [114]. The construction of the brightfield microscopes can be further simplified and made cost-effective by inkjet printing of lenses, using polydimethylsiloxane (PDMS) [115].

Smartphones can be also used as a fluorescent microscope which is an essential tool for modern biomedical diagnostics. A typical smartphone fluorescence microscope consists of an excitation light source (LED or laser diode), lenses and an emission filter. The wavelength of the exciting light is shorter than the emitted light and is filtered out before the detection of the fluorescent photon. Smartphones have also been used for phase-contrast imaging [116–118]. Similarly, spectroscopic measurements were performed by integrating dispersive elements, e.g., diffraction gratings, and pinholes or optical fibers [119].

Over the past decade, these smartphone-based technologies have been widely used for the detection of different pathogens [57,120–130]. A unique fluorescence-based approach of combining quantum dot barcode technology with smartphones was used to detect HIV and Hepatitis-B virus [131].

A schematic of the assay is shown in Figure 5. The patient's sample was first added to a chip, after amplifying the genetic target via an isothermal amplification process. The chip was composed of microbeads barcoded by different colored quantum dots. These quantum dots in turn were coated with specific recognition molecules (capture DNA). The target DNA in the sample binds to their respective capture DNA present on the microbeads. A fluorescently labeled secondary targeting agent (detection DNA) was then introduced, which specifically binds to the other end of the target DNA, thus forming a sandwich structure. The color of the fluorescence label of the detection DNA was different from the quantum dots and thus their co-localization confirmed the presence of the target viral DNA. One of the major advantages of this barcode technology is the ability to simultaneously detect multiple different viruses. A specially designed smartphone (Apple iPhone 4S) attachment containing two diode lasers (for chip illumination), a set of excitation and emission (bandpass) filters along with an objective and an eyepiece, were used as the barcode readout. The limit of detection of this assay was ~1000 viral copies/mL.

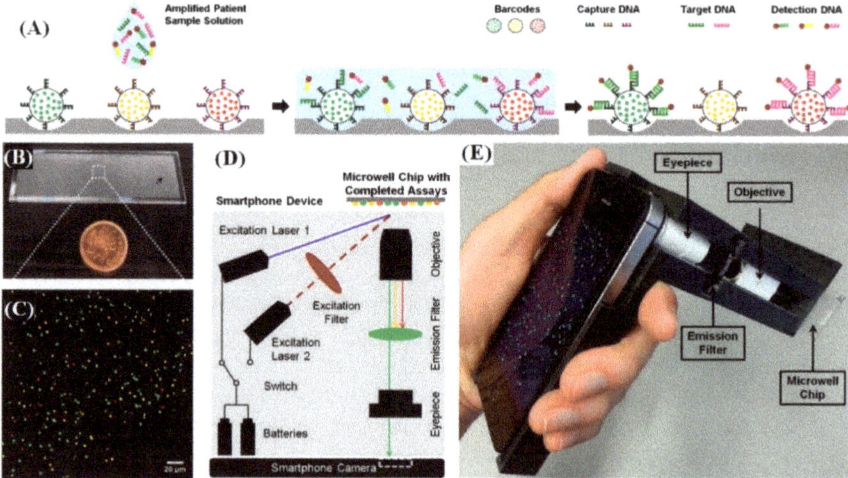

Figure 5. (**A**) Schematic of the fluorescence assay for detecting multiple pathogens, using a smartphone: The sample was added to a chip coated with microbeads, which are optically barcoded by quantum dots and are coated with bio-recognition element to capture a specific target molecule; (**B**) Photograph of the microwell chip containing different barcodes in each well; (**C**) Fluorescence image of the different quantum dot barcode array (Scale bar—20 μm); (**D**) Schematic of the smartphone device. Two excitation sources were used to excite the quantum dot barcoded chip independently. The optical emission is collected by a set of objective and eyepiece lenses and filtered using a long-pass filter and then imaged, using a smartphone camera; (**E**) Photograph of the smartphone device incorporated with the microwell chip. Used with permission, from Reference [131].

Another smartphone-based technology capable of detecting multiple mosquito-borne viruses, i.e., Zika (ZIKV), chikungunya (CHIKV), and dengue (DENV), was developed using a loop-mediated isothermal amplification (LAMP) box [132]. The box consisted of a heating module, a housing module for the assay, a detector, and an analyzer unit to interpret the data. A heating module was used to warm the sample to about 70 °C. A dry shelf-stable assay was used containing a primer and dyes, that can be rehydrated with water and amplification buffer, prior to the assay. Furthermore, one primer was used to test three strains of ZIKV, and different primers were used for both CHIKV and DENV. Different human samples, such as urine, saliva, and blood, were spiked with the virus and then tested at various concentrations. The change in fluorescence signal due to the presence of the target virus was detected by irradiating the sample, using a 3-watt RGB LED coupled to an RGB multiband pass filter.

The fluorescence images were captured by using a smartphone, and the data were analyzed, using a custom-built application (app). Furthermore, the app was also used to control the laser and heating module via a Bluetooth microcontroller.

Recently, a smartphone-based device coupled to a microfluidic chip was used for the detection of human Kaposi's sarcoma herpesvirus 8 (KSHV) [133]. The microfluidic chip containing gold nanoparticles was coated with oligonucleotides (specific to KSHV), that aggregates in the presence of the target virus. The level of nanoparticle aggregation is proportional to the viral load and results in a change in its optical (plasmonic) properties. This change was detected by irradiating the microfluidic channel with a 520 nm LED (peak SPR wavelength) and monitoring the change in voltage across an optical sensor (photocell) placed opposite to the chip. A direct correlation was observed between the voltage drop and the optical density of the sample. The data were collected and analyzed, using a smartphone. The operating range of this device was between 500 pM and 1 μM.

In another approach, *Mycobacterium tuberculosis* (MTBC), known to cause tuberculosis in human, was detected by using a paper-based assay and smartphone camera [134]. An array of wells was fabricated, using wax-based ink and impregnated with magnesium chloride ($MgCl_2$). A solution containing gold nanoparticles functionalized with thiol-modified ssDNA oligonucleotides [135], complementary to the RNA polymerase β-subunit gene of (MTBC), was used for the detection of the MTBC. The presence of MTBC would prevent the aggregation of the gold nanoparticles due to the presence of $MgCl_2$, thus preserving the red color. This change in color due to the absence of the bacteria was quantified by imaging the wells, using a mobile camera, and performing a simple RGB analysis on the images. The limit of detection of this device was reported to be 10 μg/mL MTBC sample DNA [134]. In another study, a smartphone-based fluorescence imaging platform was developed for the detection of *E. coli* in liquid samples using a sandwich immunoassay [136]. For this purpose, glass capillaries coated with anti-*E. coli* O157:H7 antibody were used to capture the *E. coli* particles in a contaminated sample. A secondary anti-*E. coli* antibody conjugated to biotin was then added in order to make a sandwich structure. The final step involved introducing streptavidin-conjugated quantum dots, which would bind to the biotin, tagged with the secondary antibodies, thereby labeling them. The capillary tubes were used to deliver the liquid into the imaging volume and served as a waveguide for the excitation light. The fluorescence emission from the quantum dots, attached to *E. coli* particles, were imaged and quantified using a cost-effective and lightweight smart-phone microscope. The detection limit of this platform was measured to be ~5 to 10 CFU/mL.

5. Lensless Digital Holographic Imaging

Lensless holographic imaging is another portable imaging modality that has gained prominence in the last decade due to its low cost, compactness, and wide field-of-view, which increases the throughput [93,137–144]. A lensless imaging platform is relatively simple and can be built using inexpensive light sources, e.g., LED, and a complementary metal-oxide semiconductor (CMOS) imaging sensor. A partially coherent light source is used to illuminate the sample and the resulting in-line holograms are recorded in the imaging sensor. The holograms are formed on the imaging chip due to the interference between the scattered wave from the semi-transparent sample and the transmitted wave. These holograms are digitally backpropagated to the object plane in order to reconstruct the image of the sample. Holography, being an interferometric technique, enables the extraction of both the amplitude and phase information following reconstruction. There are several approaches to reconstruct the images. One of the most commonly used technique is the angular spectrum approach, which involves multiplying the Fourier transform of the captured hologram with a transfer function $H_{z_2}(f_x, f_y)$, and taking the inverse fourier transform of the product to recover the image [128]. This is expressed as follows [145]:

$$E_r = F^{-1}\{F\{E_i(x, y)\}H_{z_2}(f_x, f_y)\} \tag{1}$$

where E_r is the reconstructed optical field of the object, $E_i(x, y)$ is the captured hologram and $H_{z_2}(f_x, f_y)$ is the transfer function of the free space ($n = 1$). The transfer function is defined as [145]:

$$H_{z_2}(f_x, f_y) = \begin{cases} e^{ikz_2\sqrt{1-\left(\frac{2\pi f_x}{k}\right)^2-\left(\frac{2\pi f_y}{k}\right)^2}}, & f_x^2+f_y^2 < \frac{1}{\lambda^2} \\ 0, & f_x^2+f_y^2 \geq \frac{1}{\lambda^2} \end{cases} \tag{2}$$

Here, λ is the wavelength of the light, $k = \frac{2\pi}{\lambda}$; f_x and f_y are spatial frequencies; and z_2 is the sample to sensor distance. The sample to sensor distance is kept small (<1 mm), thereby leading to unit magnification. Thus, the field of view of this imaging system is quite large, compared to a conventional lens-based imaging system. The resolution of this type of imaging system is typically limited by the degree of coherence and pixel size of the CMOS sensor, but using different super-resolution techniques it was possible to achieve resolution sub-diffraction limited resolution (~250 nm) [146].

The small size and low cost of these devices make them ideal for POC applications and several holographic imaging devices have been used for the detection of different types of viruses and bacteria [147–150]. In one such example, a digital lensless microscope was used to detect the Herpes Simplex Virus (HSV-1), using a microparticle clustering assay [151]. In this assay, silica microparticles (~2 μm) coated with HSV-1 antibodies were mixed with the viral particles in solution and imaged by using a holographic microscope, as shown in Figure 6. The presence of the virus caused the microparticles to aggregate, and the level of aggregation was used as the metric to infer the presence and concentration of the virus in the sample solution. Deep learning approaches were used for image reconstruction and analysis, which yielded a limit of detection as low as ~5 viral copies/μL (i.e., ~25 copies/test).

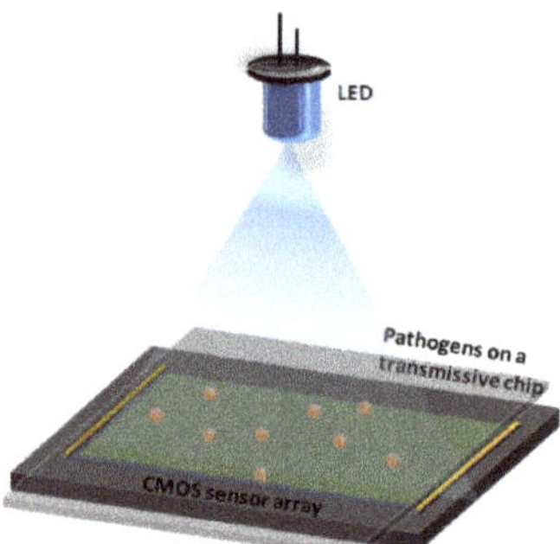

Figure 6. Schematic of a lensless digital holographic imaging system. A simple imaging system consists of a light source, complementary metal-oxide semiconductor (CMOS) sensor array, and a semi-transparent chip/substrate containing the sample.

In another example, a chip for the capture of HSV-1 virus was prepared by first functionalizing the glass with silane-PEG-biotin and then adding streptavidin to it, as shown in Figure 7 [152]. Non-specific binding was eliminated by coating the glass coverslip with m-PEG-silane. The virus sample was then incubated with biotinylated antibodies specific to HSV-1. The virus-antibody-biotin was then added to the chip, in order to capture them to the surface via the biotin–avidin bond. This chip was then imaged using a lensless microscope, with pixel super-resolution capability. This was achieved by illuminating the sample sequentially with 20 different LEDs, in order to record holograms with sub-pixel shifts. These sub-pixels shifted holograms were then used to synthesize a high-resolution hologram, which was reconstructed to obtain the phase and amplitude images of the virus. Another salient feature is the use of nanolens in order to amplify the optical signature of the viral particles. Poly-ethylene glycol (PEG-400) vapor was condensed onto the chip, resulting in the formation of drop-like structures selectively around the viral particles that act as lenses. The peak phase value of the reconstructed images of the virus and nanolens was used to estimate the size of the particles and thus confirm the presence of the HSV-1 virus, which has a size of ~150–200 nm. An automated program was used to count the number of viral particles by digitally filtering out the particles with sizes outside 150–200 nm. A limit of detection of 120 viral particles per test over a field of view of ~30 mm^2 was reported. Another approach of detecting pathogens was demonstrated by using an acoustically actuated holographic microscope, which facilitated the detection of virus (HSV-1) and bacteria (*Staphylococcus aureus*) in solution. In this approach, an acoustic transducer was coupled to a chip containing the pathogens [153]. The interdigitated transducer was used to generate surface acoustic waves which interacted with the chip to generate dispersive Lamb-type guided waves. This energy was coupled onto the liquid layer containing the pathogen and led to the formation of standing waves. The formation of the standing waves resulted in the displacement of the fluid from the antinode region, thereby exposing the pathogens and creating localized lens like liquid menisci around it. These lens-like structures (menisci) enabled the detection of *Staphylococcus aureus* bacteria and HSV-1 virus, by imaging them, using a low-cost portable holographic microscope.

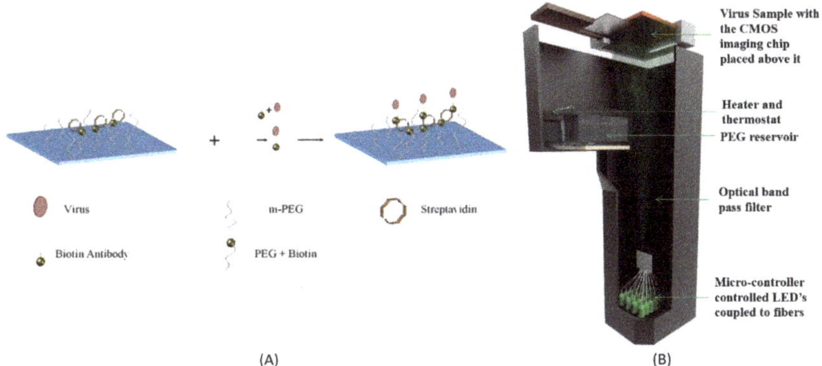

Figure 7. (**A**) Schematic of the HSV-1 capture assay on a specially prepared chip. (**B**) Schematic of the portable lensless microscope with pixel super-resolution capability. The device weighs less than 500 gm and is about 25 cm in height [152]. Lensless microscopy was also used to detect *Staphylococcus aureus* directly in a contact lens [154]. The contact lenses were coated with multiple layers polyelectrolytes that enables the immobilization of antibody specific to the *S. aureus* onto them. Simulated experiments were performed by incubating the antibody-coated contact lens with artificial tear fluid containing the bacteria. This was followed by the addition of a secondary antibody-coated polystyrene microparticle (5 μm), which resulted in the formation of a sandwich structure. A portable lensless microscope was used to directly image and quantify the number of microparticles present on the curved surface of the contact lens. Up to 16 bacteria/μL could be detected by using this method.

6. Conclusions

In this review, we described some recent point-of-care technologies incorporating plasmonics, microfluidics, smartphone imagers, and lensless microscopes for simple, sensitive, rapid 'on-site' detection of pathogens (summarized in Table 1). Several examples covering a wide range of techniques such as immunoassays (ELISA, fluorescence, etc.) and nucleic acid amplification were discussed. Although the POC devices have been able to overcome some of the major drawbacks associated with conventional diagnostic technologies, particularly in terms of cost, throughput, and portability, there are still ways to go. A huge amount of effort needs to be dedicated in order to improve their sensitivity, specificity, ease of use, and storage, which will facilitate the use of these diagnostic techniques everywhere around the globe daily. These advanced POC devices hold the potential to revolutionize the diagnosis of the viral and bacterial pathogens, especially in resource-limited settings, thereby saving countless more lives.

Table 1. A list of commonly used point-of-care (POC) technologies for the detection of some of the highly infectious bacterial and viral pathogens.

Pathogens	Detection Platform	Detection Device	Type of Assay Used	References
SARS-CoV-2	LFIA	Visual read	RT-LAMP and CRISPR	[59]
H1N1	Microfluidics	Amperometry	Electrochemical	[60]
Zika virus	LFIA	Smartphone	Fluorescent Immunoassay	[61]
	Microfluidics	Smartphone	ELISA	[62]
	Microfluidics (Paper)	Smartphone	RT-LAMP	[55]
	Plasmonics	Spectral shift	Immunoassay	[107]
Zika, Dengue and Chikungunya	Reaction tubes	Smartphone	LAMP	[132]
HIV	LFIA	Smartphone	RT-LAMP	[69]
	Microfluidics	Electric sensing	Immunoassay	[71]
	Plasmonic	Spectral shift	Immunoassay	[106]
HIV and Hep. B	Barcoded chip	Smartphone	Isothermal amplification	[131]
H7N9	Plasmonics		Immunoassay	[104]
Kaposi sarcoma herpesvirus 8	Microfluidics	Smartphone	Nanoparticle aggregation	[133]
HSV1	Glass Chip	Lensless Holographic microscope	Microparticle clustering	[151]
	Surface functionalized glass Chip	Lensless Holographic microscope	Size-based Immunoassay	[152]
S. aureus	Contact Lens	Holographic microscope	Immunoassay	[154]
E. coli	Paper microfluidic	Smartphone	Nanoparticle aggregation	[71]
	Plasmonics	CCD	Immunoassay	[108]
	Glass capillaries	Smartphone	Sandwich Immunoassay	[136]
M. tuberculosis	Microfluidic	Amperometry	Electrochemical ELISA	[74]
	Plasmonics	Optical Sensor Array	Immunoassay	[109]
	Paper/plasmonics	Smartphone	Nanoparticle aggregation	[134]

Author Contributions: Conceptualization, A.R.; methodology, P.N., A.K., S.K.D., Z.J.K., and A.R.; writing—original draft preparation, P.N., A.K., S.K.D., Z.J.K., and A.R.; writing—review and editing, P.N. and A.R.; supervision, A.R.; All authors have read and agreed to the published version of the manuscript.

Funding: This research was funded by the University of Toledo startup funds (2019).

Acknowledgments: The authors would like to thank Ms. Valentina Valentina Campos Yanez for the schematic of the smartphone reader in Figure 1.

Conflicts of Interest: The authors declare no conflict of interest.

References

1. Cook, A.H.; Cohen, D.B. Pandemic Disease: A Past and Future Challenge to Governance in the United States. *Rev. Policy Res.* **2008**, *25*, 449–471. [CrossRef]
2. Balkhair, A.A. COVID-19 Pandemic: A New Chapter in the History of Infectious Diseases. *Oman Med. J.* **2020**, e123. [CrossRef] [PubMed]
3. National Academies of Sciences, Engineering, and Medicine. *Global Health and the Future Role of the United States*; National Academies Press: Washington, DC, USA, 2017. [CrossRef]
4. Huremović, D. Brief History of Pandemics (Pandemics Throughout History). In *Psychiatry of Pandemics*; Springer: Cham, Switzerland, 2019; pp. 7–35. [CrossRef]
5. van Seventer, J.M.; Hochberg, N.S. Principles of Infectious Diseases: Transmission, Diagnosis, Prevention, and Control. *Int. Encycl. Public Health* **2017**, 22–39. [CrossRef]
6. Fauci, A.S.; Touchette, N.A.; Folkers, G.K. Emerging Infectious Diseases: A 10-Year Perspective from the National Institute of Allergy and Infectious Diseases. *Emerg. Infect. Dis.* **2005**, *11*, 519–525. [CrossRef]
7. Lemon, S.M.; Hamburg, M.A.; Sparling, P.F.; Choffnes, E.R.; Mack, A.; Rapporteurs Institute of Medicine (US) Forum on Microbial Threats. *Global Infectious Disease Surveillance and Detection: Assessing the Challenges—Finding Solutions, Workshop Summary*; National Academies Press (US): Washington, DC; USA, 2007. [CrossRef]
8. Tran, B.X.; Phan, H.T.; Latkin, C.A.; Nguyen, H.L.T.; Hoang, C.L.; Ho, C.S.H.; Ho, R.C.M. Understanding Global HIV Stigma and Discrimination: Are Contextual Factors Sufficiently Studied? (GAP(RESEARCH)). *Int. J. Environ. Res. Public Health* **2019**, *16*, 1899. [CrossRef] [PubMed]
9. Taubenberger, J.K.; Morens, D.M. The Pathology of Influenza Virus Infections. *Annu. Rev. Pathol.* **2008**, *3*, 499–522. [CrossRef] [PubMed]
10. Morens, D.M.; Taubenberger, J.K.; Harvey, H.A.; Memoli, M.J. The 1918 Influenza Pandemic: Lessons for 2009 and the Future. *Crit. Care Med.* **2010**, *38* (Suppl. S4), e10–e20. [CrossRef] [PubMed]
11. Palacios Cruz, M.; Santos, E.; Velázquez Cervantes, M.A.; León Juárez, M. COVID-19, a Worldwide Public Health Emergency TT—COVID-19, Una Emergencia de Salud Pública Mundial. *Rev. Clin. Esp.* **2020**. [CrossRef]
12. CDC COVID Data Tracker. Available online: https://covid.cdc.gov/covid-data-tracker/ (accessed on 20 August 2020).
13. Gharehgozli, O.; Nayebvali, P.; Gharehgozli, A.; Zamanian, Z. Impact of COVID-19 on the Economic Output of the US Outbreak's Epicenter. *Econ. Disasters Clim. Chang.* **2020**, *4*, 561–573. [CrossRef]
14. *Global Tuberculosis Report*; WHO: Geneva, Switzerland, 2019; ISBN 978-92-4-156571-4.
15. Al-Humadi, H.W.; Al-Saigh, R.J.; Al-Humadi, A.W. Addressing the Challenges of Tuberculosis: A Brief Historical Account. *Front. Pharmacol.* **2017**, *8*, 689. [CrossRef]
16. Blumberg, H.M.; Ernst, J.D. The Challenge of Latent TB Infection. *JAMA* **2016**, *316*, 931–933. [CrossRef]
17. Sudre, P.; ten Dam, G.; Kochi, A. Tuberculosis: A Global Overview of the Situation Today. *Bull. World Health Organ.* **1992**, *70*, 149–159.
18. Harries, A.D.; Kumar, A.M.V. Challenges and Progress with Diagnosing Pulmonary Tuberculosis in Low- and Middle-Income Countries. *Diagnostics* **2018**, *8*, 78. [CrossRef] [PubMed]
19. Fenwick, A. Waterborne Infectious Diseases—Could They Be Consigned to History? *Science* **2006**, *313*, 1077–1081. [CrossRef]
20. Shannon, M.A.; Bohn, P.W.; Elimelech, M.; Georgiadis, J.G.; Mariñas, B.J.; Mayes, A.M. Science and Technology for Water Purification in the Coming Decades. *Nature* **2008**, *452*, 301–310. [CrossRef] [PubMed]
21. Bintsis, T. Foodborne Pathogens. *AIMS Microbiol.* **2017**, *3*, 529–563. [CrossRef]

22. Michaud, C.M. Global Burden of Infectious Diseases. *Encycl. Microbiol.* **2009**, 444–454. [CrossRef]

23. Srivastava, S.; Singh, P.K.; Vatsalya, V.; Karch, R.C. Developments in the Diagnostic Techniques of Infectious Diseases: Rural and Urban Prospective. *Adv. Infect. Dis.* **2018**, *8*, 121–138. [CrossRef]

24. Desselberger, U.; Collingham, K. Molecular Techniques in the Diagnosis of Human Infectious Diseases. *Genitourin. Med.* **1990**, *66*, 313–323. [CrossRef]

25. Murray, P.R.; Masur, H. Current Approaches to the Diagnosis of Bacterial and Fungal Bloodstream Infections in the Intensive Care Unit. *Crit. Care Med.* **2012**, *40*, 3277–3282. [CrossRef]

26. Tang, Y.-W.; Schmitz, J.E.; Persing, D.H.; Stratton, C.W. Laboratory Diagnosis of COVID-19: Current Issues and Challenges. *J. Clin. Microbiol.* **2020**, *58*, e00512-20. [CrossRef] [PubMed]

27. Bloom, D.E.; Cadarette, D. Infectious Disease Threats in the Twenty-First Century: Strengthening the Global Response. *Front. Immunol.* **2019**, *10*, 549. [CrossRef] [PubMed]

28. St John, A.; Price, C.P. Existing and Emerging Technologies for Point-of-Care Testing. *Clin. Biochem. Rev.* **2014**, *35*, 155–167.

29. Christodouleas, D.C.; Kaur, B.; Chorti, P. From Point-of-Care Testing to EHealth Diagnostic Devices (EDiagnostics). *ACS Cent. Sci.* **2018**, *4*, 1600–1616. [CrossRef] [PubMed]

30. Vashist, S.K. Point-of-Care Diagnostics: Recent Advances and Trends. *Biosensors* **2017**, *7*, 62. [CrossRef] [PubMed]

31. Alawsi, T.; Al-Bawi, Z. A Review of Smartphone Point-of-Care Adapter Design. *Eng. Rep.* **2019**, *1*, e12039. [CrossRef]

32. Moon, S.; Keles, H.O.; Kim, Y.-G.; Kuritzkes, D.; Demirci, U. Lensless Imaging for Point-of-Care Testing. In Proceedings of the Annual International Conference of the IEEE Engineering in Medicine and Biology Society, Minneapolis, MN, USA, 3–6 September 2009; pp. 6376–6379. [CrossRef]

33. Tokel, O.; Inci, F.; Demirci, U. Advances in Plasmonic Technologies for Point of Care Applications. *Chem. Rev.* **2014**, *114*, 5728–5752. [CrossRef]

34. Mejía-Salazar, J.R.; Cruz, K.R.; Vásques, E.M.M.; de Oliveira, O.N. Microfluidic Point-of-Care Devices: New Trends and Future Prospects for Ehealth Diagnostics. *Sensors* **2020**, *20*, 1951. [CrossRef]

35. Chin, C.D.; Laksanasopin, T.; Cheung, Y.K.; Steinmiller, D.; Linder, V.; Parsa, H.; Wang, J.; Moore, H.; Rouse, R.; Umviligihozo, G.; et al. Microfluidics-Based Diagnostics of Infectious Diseases in the Developing World. *Nat. Med.* **2011**, *17*, 1015–1019. [CrossRef]

36. Gale, B.K.; Jafek, A.R.; Lambert, C.J.; Goenner, B.L.; Moghimifam, H.; Nze, U.C.; Kamarapu, S.K. A Review of Current Methods in Microfluidic Device Fabrication and Future Commercialization Prospects. *Inventions* **2018**, *3*, 60. [CrossRef]

37. Pandey, C.M.; Augustine, S.; Kumar, S.; Kumar, S.; Nara, S.; Srivastava, S.; Malhotra, B.D. Microfluidics Based Point-of-Care Diagnostics. *Biotechnol. J.* **2018**, *13*, 1700047. [CrossRef] [PubMed]

38. Sia, S.K.; Kricka, L.J. Microfluidics and Point-of-Care Testing. *Lab Chip* **2008**, *8*, 1982–1983. [CrossRef] [PubMed]

39. Sher, M.; Zhuang, R.; Demirci, U.; Asghar, W. Paper-Based Analytical Devices for Clinical Diagnosis: Recent Advances in the Fabrication Techniques and Sensing Mechanisms. *Expert Rev. Mol. Diagn.* **2017**, *17*, 351–366. [CrossRef] [PubMed]

40. Nilghaz, A.; Liu, X.; Ma, L.; Huang, Q.; Lu, X. Development of Fabric-Based Microfluidic Devices by Wax Printing. *Cellulose* **2019**, *26*, 3589–3599. [CrossRef]

41. Ren, K.; Zhou, J.; Wu, H. Materials for Microfluidic Chip Fabrication. *Acc. Chem. Res.* **2013**, *46*, 2396–2406. [CrossRef]

42. Becker, H.; Locascio, L.E. Polymer Microfluidic Devices. *Talanta* **2002**, *56*, 267–287. [CrossRef]

43. Hristov, D.R.; Rodriguez-Quijada, C.; Gomez-Marquez, J.; Hamad-Schifferli, K. Designing Paper-Based Immunoassays for Biomedical Applications. *Sensors* **2019**, *19*, 554. [CrossRef]

44. Kasetsirikul, S.; Shiddiky, M.J.A.; Nguyen, N.-T. Challenges and Perspectives in the Development of Paper-Based Lateral Flow Assays. *Microfluid. Nanofluid.* **2020**, *24*, 17. [CrossRef]

45. Joung, H.-A.; Ballard, Z.S.; Ma, A.; Tseng, D.K.; Teshome, H.; Burakowski, S.; Garner, O.B.; Di Carlo, D.; Ozcan, A. Paper-Based Multiplexed Vertical Flow Assay for Point-of-Care Testing. *Lab Chip* **2019**, *19*, 1027–1034. [CrossRef]

46. Ariffin, N.; Yusof, N.A.; Abdullah, J.; Abd Rahman, S.F.; Ahmad Raston, N.H.; Kusnin, N.; Suraiya, S. Lateral Flow Immunoassay for Naked Eye Detection of *Mycobacterium tuberculosis*. *J. Sens.* **2020**, *2020*. [CrossRef]

47. Zhu, H.; Fohlerová, Z.; Pekárek, J.; Basova, E.; Neužil, P. Recent Advances in Lab-on-a-Chip Technologies for Viral Diagnosis. *Biosens. Bioelectron.* **2020**, *153*, 112041. [CrossRef] [PubMed]
48. Nguyen, T.; Chidambara, V.A.; Andreasen, S.Z.; Golabi, M.; Huynh, V.N.; Linh, Q.T.; Bang, D.D.; Wolff, A. Point-of-Care Devices for Pathogen Detections: The Three Most Important Factors to Realise towards Commercialization. *TrAC Trends Anal. Chem.* **2020**, *131*, 116004. [CrossRef]
49. Tsao, C.-W. Polymer Microfluidics: Simple, Low-Cost Fabrication Process Bridging Academic Lab Research to Commercialized Production. *Micromachines* **2016**, *7*, 225. [CrossRef] [PubMed]
50. Cao, Q.; Mahalanabis, M.; Chang, J.; Carey, B.; Hsieh, C.; Stanley, A.; Odell, C.A.; Mitchell, P.; Feldman, J.; Pollock, N.R.; et al. Microfluidic Chip for Molecular Amplification of Influenza A RNA in Human Respiratory Specimens. *PLoS ONE* **2012**, *7*, e33176. [CrossRef] [PubMed]
51. Qin, P.; Park, M.; Alfson, K.J.; Tamhankar, M.; Carrion, R.; Patterson, J.L.; Griffiths, A.; He, Q.; Yildiz, A.; Mathies, R.; et al. Rapid and Fully Microfluidic Ebola Virus Detection with CRISPR-Cas13a. *ACS Sens.* **2019**, *4*, 1048–1054. [CrossRef]
52. Zhao, C.; Liu, X. A Portable Paper-Based Microfluidic Platform for Multiplexed Electrochemical Detection of Human Immunodeficiency Virus and Hepatitis C Virus Antibodies in Serum. *Biomicrofluidics* **2016**, *10*, 24119. [CrossRef] [PubMed]
53. Zhao, X.; Li, M.; Liu, Y. Microfluidic-Based Approaches for Foodborne Pathogen Detection. *Microorganisms* **2019**, *7*, 381. [CrossRef] [PubMed]
54. Mauk, M.; Song, J.; Bau, H.H.; Gross, R.; Bushman, F.D.; Collman, R.G.; Liu, C. Miniaturized Devices for Point of Care Molecular Detection of HIV. *Lab Chip* **2017**, *17*, 382–394. [CrossRef] [PubMed]
55. Kaarj, K.; Akarapipad, P.; Yoon, J.-Y. Simpler, Faster, and Sensitive Zika Virus Assay Using Smartphone Detection of Loop-Mediated Isothermal Amplification on Paper Microfluidic Chips. *Sci. Rep.* **2018**, *8*, 12438. [CrossRef]
56. Wang, S.; Inci, F.; De Libero, G.; Singhal, A.; Demirci, U. Point-of-Care Assays for Tuberculosis: Role of Nanotechnology/Microfluidics. *Biotechnol. Adv.* **2013**, *31*, 438–449. [CrossRef]
57. Nasseri, B.; Soleimani, N.; Rabiee, N.; Kalbasi, A.; Karimi, M.; Hamblin, M.R. Point-of-Care Microfluidic Devices for Pathogen Detection. *Biosens. Bioelectron.* **2018**, *117*, 112–128. [CrossRef] [PubMed]
58. NIH Delivering New COVID-19 Testing Technologies to Meet U.S. Demand. Available online: https://www.nih.gov/news-events/news-releases/nih-delivering-new-covid-19-testing-technologies-meet-us-demand (accessed on 31 July 2020).
59. Joung, J.; Ladha, A.; Saito, M.; Segel, M.; Bruneau, R.; Huang, M.W.; Kim, N.G.; Yu, X.; Li, J.; Walker, B.D.; et al. Point-of-Care Testing for COVID-19 Using SHERLOCK Diagnostics. *medRxiv Prep. Serv. Health Sci.* **2020**, 20091231. [CrossRef]
60. Singh, R.; Hong, S.; Jang, J. Label-Free Detection of Influenza Viruses Using a Reduced Graphene Oxide-Based Electrochemical Immunosensor Integrated with a Microfluidic Platform. *Sci. Rep.* **2017**, *7*, 42771. [CrossRef]
61. Rong, Z.; Wang, Q.; Sun, N.; Jia, X.; Wang, K.; Xiao, R.; Wang, S. Smartphone-Based Fluorescent Lateral Flow Immunoassay Platform for Highly Sensitive Point-of-Care Detection of Zika Virus Nonstructural Protein 1. *Anal. Chim. Acta* **2019**, *1055*, 140–147. [CrossRef] [PubMed]
62. Kabir, M.A.; Zilouchian, H.; Sher, M.; Asghar, W. Development of a Flow-Free Automated Colorimetric Detection Assay Integrated with Smartphone for Zika NS1. *Diagnostics* **2020**, *10*, 42. [CrossRef]
63. Da Silva, S.J.R.; Pardee, K.; Pena, L. Loop-Mediated Isothermal Amplification (LAMP) for the Diagnosis of Zika Virus: A Review. *Viruses* **2019**, *12*, 19. [CrossRef] [PubMed]
64. Da Silva, S.J.R.; Paiva, M.H.S.; Guedes, D.R.D.; Krokovsky, L.; de Melo, F.L.; da Silva, M.A.L.; da Silva, A.; Ayres, C.F.J.; Pena, L.J. Development and Validation of Reverse Transcription Loop-Mediated Isothermal Amplification (RT-LAMP) for Rapid Detection of ZIKV in Mosquito Samples from Brazil. *Sci. Rep.* **2019**, *9*, 4494. [CrossRef] [PubMed]
65. Martin, C.A.; Keren, D.F. Comparison of Murex Single-Use Diagnostic System with Traditional Enzyme Immunoassay for Detection of Exposure to Human Immunodeficiency Virus. *Clin. Diagn. Lab. Immunol.* **2002**, *9*, 187–189. [CrossRef]
66. Nandi, S.; Maity, S.; Bhunia, S.C.; Saha, M.K. Comparative Assessment of Commercial ELISA Kits for Detection of HIV in India. *BMC Res. Notes* **2014**, *7*, 436. [CrossRef]
67. US Foods. *OraQuick In-Home HIV Test*; US Foods: Rosemont, IL, USA, 2020.

68. Gueudin, M.; Leoz, M.; Lemée, V.; De Oliveira, F.; Vessière, A.; Kfutwah, A.; Plantier, J.-C. A New Real-Time Quantitative PCR for Diagnosis and Monitoring of HIV-1 Group O Infection. *J. Clin. Microbiol.* **2012**, *50*, 831–836. [CrossRef]

69. Phillips, E.A.; Moehling, T.J.; Ejendal, K.; Hoilett, O.S.; Byers, K.M.; Basing, L.A.; Jankowski, L.A.; Bennett, J.B.; Lin, L.K.; Stanciu, L.A.; et al. Microfluidic Rapid and Autonomous Analytical Device (MicroRAAD) to Detect HIV from Whole Blood Samples. *Lab Chip* **2019**, *19*, 3375–3386. [CrossRef]

70. Pilcher, C.D.; Joaki, G.; Hoffman, I.F.; Martinson, F.E.; Mapanje, C.; Stewart, P.W.; Powers, K.A.; Galvin, S.; Chilongozi, D.; Gama, S.; et al. Amplified Transmission of HIV-1: Comparison of HIV-1 Concentrations in Semen and Blood during Acute and Chronic Infection. *AIDS* **2007**, *21*, 1723–1730. [CrossRef] [PubMed]

71. Shafiee, H.; Asghar, W.; Inci, F.; Yuksekkaya, M.; Jahangir, M.; Zhang, M.H.; Durmus, N.G.; Gurkan, U.A.; Kuritzkes, D.R.; Demirci, U. Paper and Flexible Substrates as Materials for Biosensing Platforms to Detect Multiple Biotargets. *Sci. Rep.* **2015**, *5*, 8719. [CrossRef] [PubMed]

72. Pai, M.; Denkinger, C.M.; Kik, S.V.; Rangaka, M.X.; Zwerling, A.; Oxlade, O.; Metcalfe, J.Z.; Cattamanchi, A.; Dowdy, D.W.; Dheda, K.; et al. Gamma Interferon Release Assays for Detection of *Mycobacterium tuberculosis* Infection. *Clin. Microbiol. Rev.* **2014**, *27*, 3–20. [CrossRef]

73. Lalvani, A.; Pareek, M. Interferon Gamma Release Assays: Principles and Practice. *Enferm. Infecc. Microbiol. Clin.* **2010**, *28*, 245–252. [CrossRef] [PubMed]

74. Evans, D.; Papadimitriou, K.I.; Greathead, L.; Vasilakis, N.; Pantelidis, P.; Kelleher, P.; Morgan, H.; Prodromakis, T. An Assay System for Point-of-Care Diagnosis of Tuberculosis Using Commercially Manufactured PCB Technology. *Sci. Rep.* **2017**, *7*, 685. [CrossRef]

75. Wang, S.; Shan, X.; Patel, U.; Huang, X.; Lu, J.; Li, J.; Tao, N. Label-Free Imaging, Detection, and Mass Measurement of Single Viruses by Surface Plasmon Resonance. *Proc. Natl. Acad. Sci. USA* **2010**, *107*, 16028–16032. [CrossRef]

76. Singh, P. Surface Plasmon Resonance: A Boon for Viral Diagnostics. *Ref. Modul. Life Sci.* **2017**. [CrossRef]

77. Piliarik, M.; Vaisocherová, H.; Homola, J. Surface Plasmon Resonance Biosensing. *Methods Mol. Biol.* **2009**, *503*, 65–88. [CrossRef]

78. Nguyen, H.H.; Park, J.; Kang, S.; Kim, M. Surface Plasmon Resonance: A Versatile Technique for Biosensor Applications. *Sensors* **2015**, *15*, 10481–10510. [CrossRef]

79. Choi, I.; Choi, Y. Plasmonic Nanosensors: Review and Prospect. *IEEE J. Sel. Top. Quantum Electron.* **2012**, *18*, 1110–1121. [CrossRef]

80. Soler, M.; Huertas, C.S.; Lechuga, L.M. Label-Free Plasmonic Biosensors for Point-of-Care Diagnostics: A Review. *Expert Rev. Mol. Diagn.* **2019**, *19*, 71–81. [CrossRef]

81. Pilot, R.; Signorini, R.; Durante, C.; Orian, L.; Bhamidipati, M.; Fabris, L. A Review on Surface-Enhanced Raman Scattering. *Biosensors* **2019**, *9*, 57. [CrossRef] [PubMed]

82. Langer, J.; Jimenez de Aberasturi, D.; Aizpurua, J.; Alvarez-Puebla, R.A.; Auguié, B.; Baumberg, J.J.; Bazan, G.C.; Bell, S.E.J.; Boisen, A.; Brolo, A.G.; et al. Present and Future of Surface-Enhanced Raman Scattering. *ACS Nano* **2020**, *14*, 28–117. [CrossRef] [PubMed]

83. Li, Z.; Leustean, L.; Inci, F.; Zheng, M.; Demirci, U.; Wang, S. Plasmonic-Based Platforms for Diagnosis of Infectious Diseases at the Point-of-Care. *Biotechnol. Adv.* **2019**, *37*, 107440. [CrossRef]

84. Granger, J.H.; Schlotter, N.E.; Crawford, A.C.; Porter, M.D. Prospects for Point-of-Care Pathogen Diagnostics Using Surface-Enhanced Raman Scattering (SERS). *Chem. Soc. Rev.* **2016**, *45*, 3865–3882. [CrossRef]

85. Marks, H.; Schechinger, M.; Garza, J.; Locke, A.; Coté, G. Surface Enhanced Raman Spectroscopy (SERS) for in Vitro Diagnostic Testing at the Point of Care. *Nanophotonics* **2017**, *6*, 681–701. [CrossRef]

86. Li, B.; Singer, N.G.; Yeni, Y.N.; Haggins, D.G.; Barnboym, E.; Oravec, D.; Lewis, S.; Akkus, O. A Point-of-Care Raman Spectroscopy-Based Device for the Diagnosis of Gout and Pseudogout: Comparison With the Clinical Standard Microscopy. *Arthritis Rheumatol.* **2016**, *68*, 1751–1757. [CrossRef] [PubMed]

87. Xu, K.; Zhou, R.; Takei, K.; Hong, M. Toward Flexible Surface-Enhanced Raman Scattering (SERS) Sensors for Point-of-Care Diagnostics. *Adv. Sci.* **2019**, *6*, 1900925. [CrossRef]

88. Masson, J.-F. Portable and Field-Deployed Surface Plasmon Resonance and Plasmonic Sensors. *Analyst* **2020**, *145*, 3776–3800. [CrossRef]

89. Tokel, O.; Yildiz, U.H.; Inci, F.; Durmus, N.G.; Ekiz, O.O.; Turker, B.; Cetin, C.; Rao, S.; Sridhar, K.; Natarajan, N.; et al. Portable Microfluidic Integrated Plasmonic Platform for Pathogen Detection. *Sci. Rep.* **2015**, *5*, 9152. [CrossRef] [PubMed]

90. Wang, D.-S.; Fan, S.-K. Microfluidic Surface Plasmon Resonance Sensors: From Principles to Point-of-Care Applications. *Sensors* **2016**, *16*, 1175. [CrossRef] [PubMed]

91. Tabassum, S.; Kumar, R. Advances in Fiber-Optic Technology for Point-of-Care Diagnosis and In Vivo Biosensing. *Adv. Mater. Technol.* **2020**, *5*, 1900792. [CrossRef]

92. Vaisocherová, H.; Faca, V.M.; Taylor, A.D.; Hanash, S.; Jiang, S. Comparative Study of SPR and ELISA Methods Based on Analysis of CD166/ALCAM Levels in Cancer and Control Human Sera. *Biosens. Bioelectron.* **2009**, *24*, 2143–2148. [CrossRef] [PubMed]

93. Coskun, A.F.; Cetin, A.E.; Galarreta, B.C.; Alvarez, D.A.; Altug, H.; Ozcan, A. Lensfree Optofluidic Plasmonic Sensor for Real-Time and Label-Free Monitoring of Molecular Binding Events over a Wide Field-of-View. *Sci. Rep.* **2014**, *4*, 6789. [CrossRef] [PubMed]

94. Harpaz, D.; Koh, B.; Marks, R.S.; Seet, R.C.S.; Abdulhalim, I.; Tok, A.I.Y. Point-of-Care Surface Plasmon Resonance Biosensor for Stroke Biomarkers NT-ProBNP and S100β Using a Functionalized Gold Chip with Specific Antibody. *Sensors* **2019**, *19*, 2533. [CrossRef]

95. Lobry, M.; Loyez, M.; Chah, K.; Hassan, E.M.; Goormaghtigh, E.; DeRosa, M.C.; Wattiez, R.; Caucheteur, C. HER2 Biosensing through SPR-Envelope Tracking in Plasmonic Optical Fiber Gratings. *Biomed. Opt. Express* **2020**, *11*, 4862–4871. [CrossRef]

96. Zeni, L.; Perri, C.; Cennamo, N.; Arcadio, F.; D'Agostino, G.; Salmona, M.; Beeg, M.; Gobbi, M. A Portable Optical-Fibre-Based Surface Plasmon Resonance Biosensor for the Detection of Therapeutic Antibodies in Human Serum. *Sci. Rep.* **2020**, *10*, 11154. [CrossRef]

97. Aruna Gandhi, M.S.; Chu, S.; Senthilnathan, K.; Babu, P.R.; Nakkeeran, K.; Li, Q. Recent Advances in Plasmonic Sensor-Based Fiber Optic Probes for Biological Applications. *Appl. Sci.* **2019**, *9*, 949. [CrossRef]

98. Sharma, P.K.; Kumar, J.S.; Singh, V.V.; Biswas, U.; Sarkar, S.S.; Alam, S.I.; Dash, P.K.; Boopathi, M.; Ganesan, K.; Jain, R. Surface Plasmon Resonance Sensing of Ebola Virus: A Biological Threat. *Anal. Bioanal. Chem.* **2020**, *412*, 4101–4112. [CrossRef]

99. Diao, W.; Tang, M.; Ding, S.; Li, X.; Cheng, W.; Mo, F.; Yan, X.; Ma, H.; Yan, Y. Highly Sensitive Surface Plasmon Resonance Biosensor for the Detection of HIV-Related DNA Based on Dynamic and Structural DNA Nanodevices. *Biosens. Bioelectron.* **2018**, *100*, 228–234. [CrossRef]

100. Takemura, K.; Adegoke, O.; Suzuki, T.; Park, E.Y. A Localized Surface Plasmon Resonance-Amplified Immunofluorescence Biosensor for Ultrasensitive and Rapid Detection of Nonstructural Protein 1 of Zika Virus. *PLoS ONE* **2019**, *14*, e0211517. [CrossRef] [PubMed]

101. Usachev, E.V.; Usacheva, O.V.; Agranovski, I.E. Surface Plasmon Resonance-Based Bacterial Aerosol Detection. *J. Appl. Microbiol.* **2014**, *117*, 1655–1662. [CrossRef] [PubMed]

102. Prabowo, B.A.; Chang, Y.-F.; Lai, H.-C.; Alom, A.; Pal, P.; Lee, Y.-Y.; Chiu, N.-F.; Hatanaka, K.; Su, L.-C.; Liu, K.-C. Rapid Screening of *Mycobacterium tuberculosis* Complex (MTBC) in Clinical Samples by a Modular Portable Biosensor. *Sens. Actuators B Chem.* **2018**, *254*, 742–748. [CrossRef]

103. Koubová, V.; Brynda, E.; Karasová, L.; Škvor, J.; Homola, J.; Dostálek, J.; Tobiška, P.; Rošický, J. Detection of Foodborne Pathogens Using Surface Plasmon Resonance Biosensors. *Sens. Actuators B Chem.* **2001**, *74*, 100–105. [CrossRef]

104. Chang, Y.F.; Wang, W.H.; Hong, Y.W.; Yuan, R.Y.; Chen, K.H.; Huang, Y.W.; Lu, P.L.; Chen, Y.H.; Chen, Y.A.; Su, L.C.; et al. Simple Strategy for Rapid and Sensitive Detection of Avian Influenza A H7N9 Virus Based on Intensity-Modulated SPR Biosensor and New Generated Antibody. *Anal. Chem.* **2018**, *90*, 1861–1869. [CrossRef]

105. Whang, K.; Lee, J.-H.; Shin, Y.; Lee, W.; Kim, Y.W.; Kim, D.; Lee, L.P.; Kang, T. Plasmonic Bacteria on a Nanoporous Mirror via Hydrodynamic Trapping for Rapid Identification of Waterborne Pathogens. *Light Sci. Appl.* **2018**, *7*, 68. [CrossRef]

106. Inci, F.; Tokel, O.; Wang, S.; Gurkan, U.A.; Tasoglu, S.; Kuritzkes, D.R.; Demirci, U. Nanoplasmonic Quantitative Detection of Intact Viruses from Unprocessed Whole Blood. *ACS Nano* **2013**, *7*, 4733–4745. [CrossRef]

107. Jiang, Q.; Chandar, Y.J.; Cao, S.; Kharasch, E.D.; Singamaneni, S.; Morrissey, J.J. Rapid, Point-of-Care, Paper-Based Plasmonic Biosensor for Zika Virus Diagnosis. *Adv. Biosyst.* **2017**, *1*, e1700096. [CrossRef]

108. Wang, S.; Xie, J.; Jiang, M.; Chang, K.; Chen, R.; Ma, L.; Zhu, J.; Guo, Q.; Sun, H.; Hu, J. The Development of a Portable SPR Bioanalyzer for Sensitive Detection of *Escherichia coli* O157:H7. *Sensors* **2016**, *16*, 1856. [CrossRef]

109. Trzaskowski, M.; Ciach, T. Corrigendum for SPR System for On-Site Detection of Biological Warfare. *Curr. Anal. Chem.* **2018**, *14*, 292. [CrossRef]

110. Qiu, G.; Gai, Z.; Tao, Y.; Schmitt, J.; Kullak-Ublick, G.A.; Wang, J. Dual-Functional Plasmonic Photothermal Biosensors for Highly Accurate Severe Acute Respiratory Syndrome Coronavirus 2 Detection. *ACS Nano* **2020**, *14*, 5268–5277. [CrossRef] [PubMed]

111. Ray, A.; Esparza, S.; Wu, D.; Hanudel, M.R.; Joung, H.-A.; Gales, B.; Tseng, D.; Salusky, I.B.; Ozcan, A. Measurement of Serum Phosphate Levels Using a Mobile Sensor. *Analyst* **2020**, *145*, 1841–1848. [CrossRef] [PubMed]

112. Smith, Z.J.; Chu, K.; Espenson, A.R.; Rahimzadeh, M.; Gryshuk, A.; Molinaro, M.; Dwyre, D.M.; Lane, S.; Matthews, D.; Wachsmann-Hogiu, S. Cell-Phone-Based Platform for Biomedical Device Development and Education Applications. *PLoS ONE* **2011**, *6*, e17150. [CrossRef]

113. Hergemöller, T.; Laumann, D. Smartphone Magnification Attachment: Microscope or Magnifying Glass. *Phys. Teach.* **2017**, *55*, 361–364. [CrossRef]

114. Skandarajah, A.; Reber, C.D.; Switz, N.A.; Fletcher, D.A. Quantitative Imaging with a Mobile Phone Microscope. *PLoS ONE* **2014**, *9*, e96906. [CrossRef]

115. Sung, Y.-L.; Jeang, J.; Lee, C.-H.; Shih, W.-C. Fabricating Optical Lenses by Inkjet Printing and Heat-Assisted in Situ Curing of Polydimethylsiloxane for Smartphone Microscopy. *J. Biomed. Opt.* **2015**, *20*, 47005. [CrossRef]

116. Yang, Z.; Zhan, Q. Single-Shot Smartphone-Based Quantitative Phase Imaging Using a Distorted Grating. *PLoS ONE* **2016**, *11*, e0159596. [CrossRef]

117. Mudanyali, O.; Dimitrov, S.; Sikora, U.; Padmanabhan, S.; Navruz, I.; Ozcan, A. Integrated Rapid-Diagnostic-Test Reader Platform on a Cellphone. *Lab Chip* **2012**, *12*, 2678–2686. [CrossRef]

118. Tseng, D.; Mudanyali, O.; Oztoprak, C.; Isikman, S.O.; Sencan, I.; Yaglidere, O.; Ozcan, A. Lensfree Microscopy on a Cellphone. *Lab Chip* **2010**, *10*, 1787–1792. [CrossRef]

119. Hossain, M.A.; Canning, J.; Cook, K.; Jamalipour, A. Optical Fiber Smartphone Spectrometer. *Opt. Lett.* **2016**, *41*, 2237–2240. [CrossRef] [PubMed]

120. Ma, Y.-D.; Li, K.-H.; Chen, Y.-H.; Lee, Y.-M.; Chou, S.-T.; Lai, Y.-Y.; Huang, P.-C.; Ma, H.-P.; Lee, G.-B. A Sample-to-Answer, Portable Platform for Rapid Detection of Pathogens with a Smartphone Interface. *Lab Chip* **2019**, *19*, 3804–3814. [CrossRef] [PubMed]

121. Shrivastava, S.; Lee, W.-I.; Lee, N.-E. Culture-Free, Highly Sensitive, Quantitative Detection of Bacteria from Minimally Processed Samples Using Fluorescence Imaging by Smartphone. *Biosens. Bioelectron.* **2018**, *109*, 90–97. [CrossRef] [PubMed]

122. Rajendran, V.K.; Bakthavathsalam, P.; Jaffar Ali, B.M. Smartphone Based Bacterial Detection Using Biofunctionalized Fluorescent Nanoparticles. *Microchim. Acta* **2014**, *181*, 1815–1821. [CrossRef]

123. Cheng, N.; Song, Y.; Zeinhom, M.; Chang, Y.C.; Sheng, L.; Li, H.; Du, D.; Li, L.; Zhu, M.J.; Luo, Y.; et al. Nanozyme-Mediated Dual Immunoassay Integrated with Smartphone for Use in Simultaneous Detection of Pathogens. *ACS Appl. Mater. Interfaces* **2017**, *9*, 40671–40680. [CrossRef]

124. Wei, Q.; Qi, H.; Luo, W.; Tseng, D.; Ki, S.J.; Wan, Z.; Göröcs, Z.; Bentolila, L.A.; Wu, T.T.; Sun, R.; et al. Fluorescent Imaging of Single Nanoparticles and Viruses on a Smart Phone. *ACS Nano* **2013**, *7*, 9147–9155. [CrossRef]

125. Gopinath, S.C.B.; Tang, T.-H.; Chen, Y.; Citartan, M.; Lakshmipriya, T. Bacterial Detection: From Microscope to Smartphone. *Biosens. Bioelectron.* **2014**, *60*, 332–342. [CrossRef]

126. Hui, J.; Gu, Y.; Zhu, Y.; Chen, Y.; Guo, S.-J.; Tao, S.-C.; Zhang, Y.; Liu, P. Multiplex Sample-to-Answer Detection of Bacteria Using a Pipette-Actuated Capillary Array Comb with Integrated DNA Extraction, Isothermal Amplification, and Smartphone Detection. *Lab Chip* **2018**, *18*, 2854–2864. [CrossRef]

127. Sajid, M.; Osman, A.; Siddiqui, G.U.; Kim, H.B.; Kim, S.W.; Ko, J.B.; Lim, Y.K.; Choi, K.H. All-Printed Highly Sensitive 2D MoS2 Based Multi-Reagent Immunosensor for Smartphone Based Point-of-Care Diagnosis. *Sci. Rep.* **2017**, *7*, 5802. [CrossRef]

128. Aronoff-Spencer, E.; Venkatesh, A.G.; Sun, A.; Brickner, H.; Looney, D.; Hall, D.A. Detection of Hepatitis C Core Antibody by Dual-Affinity Yeast Chimera and Smartphone-Based Electrochemical Sensing. *Biosens. Bioelectron.* **2016**, *86*, 690–696. [CrossRef] [PubMed]

129. Barnes, L.; Heithoff, D.M.; Mahan, S.P.; Fox, G.N.; Zambrano, A.; Choe, J.; Fitzgibbons, L.N.; Marth, J.D.; Fried, J.C.; Soh, H.T.; et al. Smartphone-Based Pathogen Diagnosis in Urinary Sepsis Patients. *EBioMedicine* **2018**, *36*, 73–82. [CrossRef] [PubMed]

130. Ding, X.; Mauk, M.G.; Yin, K.; Kadimisetty, K.; Liu, C. Interfacing Pathogen Detection with Smartphones for Point-of-Care Applications. *Anal. Chem.* **2019**, *91*, 655–672. [CrossRef]

131. Ming, K.; Kim, J.; Biondi, M.J.; Syed, A.; Chen, K.; Lam, A.; Ostrowski, M.; Rebbapragada, A.; Feld, J.J.; Chan, W.C.W. Integrated Quantum Dot Barcode Smartphone Optical Device for Wireless Multiplexed Diagnosis of Infected Patients. *ACS Nano* **2015**, *9*, 3060–3074. [CrossRef] [PubMed]

132. Priye, A.; Bird, S.W.; Light, Y.K.; Ball, C.S.; Negrete, O.A.; Meagher, R.J. A Smartphone-Based Diagnostic Platform for Rapid Detection of Zika, Chikungunya, and Dengue Viruses. *Sci. Rep.* **2017**, *7*, 44778. [CrossRef]

133. Mancuso, M.; Cesarman, E.; Erickson, D. Detection of Kaposi's Sarcoma Associated Herpesvirus Nucleic Acids Using a Smartphone Accessory. *Lab Chip* **2014**, *14*, 3809–3816. [CrossRef] [PubMed]

134. Veigas, B.; Jacob, J.M.; Costa, M.N.; Santos, D.S.; Viveiros, M.; Inácio, J.; Martins, R.; Barquinha, P.; Fortunato, E.; Baptista, P.V. Gold on Paper–Paper Platform for Au-Nanoprobe TB Detection. *Lab Chip* **2012**, *12*, 4802–4808. [CrossRef] [PubMed]

135. Baptista, P.V.; Koziol-Montewka, M.; Paluch-Oles, J.; Doria, G.; Franco, R. Gold-Nanoparticle-Probe-Based Assay for Rapid and Direct Detection of *Mycobacterium tuberculosis* DNA in Clinical Samples. *Clin. Chem. Engl.* **2006**, 1433–1434. [CrossRef]

136. Zhu, H.; Sikora, U.; Ozcan, A. Quantum Dot Enabled Detection of *Escherichia coli* Using a Cell-Phone. *Analyst* **2012**, *137*, 2541–2544. [CrossRef]

137. Daloglu, M.U.; Ray, A.; Collazo, M.J.; Brown, C.; Tseng, D.; Chocarro-Ruiz, B.; Lechuga, L.M.; Cascio, D.; Ozcan, A. Low-Cost and Portable UV Holographic Microscope for High-Contrast Protein Crystal Imaging. *APL Photonics* **2019**, *4*, 30804. [CrossRef]

138. McLeod, E.; Wei, Q.; Ozcan, A. Democratization of Nanoscale Imaging and Sensing Tools Using Photonics. *Anal. Chem.* **2015**, *87*, 6434–6445. [CrossRef]

139. Ray, A.; Li, S.; Segura, T.; Ozcan, A. High-Throughput Quantification of Nanoparticle Degradation Using Computational Microscopy and Its Application to Drug Delivery Nanocapsules. *ACS Photonics* **2017**, *4*, 1216–1224. [CrossRef]

140. Zhang, Y.; Ouyang, M.; Ray, A.; Liu, T.; Kong, J.; Bai, B.; Kim, D.; Guziak, A.; Luo, Y.; Feizi, A.; et al. Computational Cytometer Based on Magnetically Modulated Coherent Imaging and Deep Learning. *Light Sci. Appl.* **2019**, *8*, 91. [CrossRef] [PubMed]

141. Khalid, M.A.; Ray, A.; Cohen, S.; Tassieri, M.; Demčenko, A.; Tseng, D.; Reboud, J.; Ozcan, A.; Cooper, J.M. Computational Image Analysis of Guided Acoustic Waves Enables Rheological Assessment of Sub-Nanoliter Volumes. *ACS Nano* **2019**, *13*, 11062–11069. [CrossRef] [PubMed]

142. Greenbaum, A.; Luo, W.; Su, T.-W.; Göröcs, Z.; Xue, L.; Isikman, S.O.; Coskun, A.F.; Mudanyali, O.; Ozcan, A. Imaging without Lenses: Achievements and Remaining Challenges of Wide-Field on-Chip Microscopy. *Nat. Methods* **2012**, *9*, 889–895. [CrossRef]

143. Daloglu, M.U.; Ray, A.; Gorocs, Z.; Xiong, M.; Malik, R.; Bitan, G.; McLeod, E.; Ozcan, A. Computational On-Chip Imaging of Nanoparticles and Biomolecules Using Ultraviolet Light. *Sci. Rep.* **2017**, *7*, 44157. [CrossRef]

144. Ozcan, A.; McLeod, E. Lensless Imaging and Sensing. *Annu. Rev. Biomed. Eng.* **2016**, *18*, 77–102. [CrossRef]

145. McLeod, E.; Ozcan, A. Unconventional Methods of Imaging: Computational Microscopy and Compact Implementations. *Rep. Prog. Phys.* **2016**, *79*, 76001. [CrossRef]

146. Luo, W.; Greenbaum, A.; Zhang, Y.; Ozcan, A. Synthetic Aperture-Based on-Chip Microscopy. *Light Sci. Appl.* **2015**, *4*, e261. [CrossRef]

147. Hennequin, Y.; Allier, C.P.; McLeod, E.; Mudanyali, O.; Migliozzi, D.; Ozcan, A.; Dinten, J.-M. Optical Detection and Sizing of Single Nanoparticles Using Continuous Wetting Films. *ACS Nano* **2013**, *7*, 7601–7609. [CrossRef]

148. McLeod, E.; Dincer, T.U.; Veli, M.; Ertas, Y.N.; Nguyen, C.; Luo, W.; Greenbaum, A.; Feizi, A.; Ozcan, A. High-Throughput and Label-Free Single Nanoparticle Sizing Based on Time-Resolved On-Chip Microscopy. *ACS Nano* **2015**, *9*, 3265–3273. [CrossRef]

149. Mudanyali, O.; McLeod, E.; Luo, W.; Greenbaum, A.; Coskun, A.F.; Hennequin, Y.; Allier, C.P.; Ozcan, A. Wide-Field Optical Detection of Nanoparticles Using on-Chip Microscopy and Self-Assembled Nanolenses. *Nat. Photonics* **2013**, *7*, 247–254. [CrossRef]

150. McLeod, E.; Nguyen, C.; Huang, P.; Luo, W.; Veli, M.; Ozcan, A. Tunable Vapor-Condensed Nanolenses. *ACS Nano* **2014**, *8*, 7340–7349. [CrossRef] [PubMed]

151. Wu, Y.; Ray, A.; Wei, Q.; Feizi, A.; Tong, X.; Chen, E.; Luo, Y.; Ozcan, A. Deep Learning Enables High-Throughput Analysis of Particle-Aggregation-Based Biosensors Imaged Using Holography. *ACS Photonics* **2019**, *6*, 294–301. [CrossRef]
152. Ray, A.; Daloglu, M.U.; Ho, J.; Torres, A.; Mcleod, E.; Ozcan, A. Computational Sensing of Herpes Simplex Virus Using a Cost-Effective on-Chip Microscope. *Sci. Rep.* **2017**, *7*, 4856. [CrossRef]
153. Ray, A.; Khalid, M.A.; Demčenko, A.; Daloglu, M.; Tseng, D.; Reboud, J.; Cooper, J.M.; Ozcan, A. Holographic Detection of Nanoparticles Using Acoustically Actuated Nanolenses. *Nat. Commun.* **2020**, *11*, 171. [CrossRef]
154. Veli, M.; Ozcan, A. Computational Sensing of *Staphylococcus aureus* on Contact Lenses Using 3D Imaging of Curved Surfaces and Machine Learning. *ACS Nano* **2018**, *12*, 2554–2559. [CrossRef] [PubMed]

Publisher's Note: MDPI stays neutral with regard to jurisdictional claims in published maps and institutional affiliations.

 *diagnostics*

Brief Report

Mobile Laboratory Reveals the Circulation of Dengue Virus Serotype I of Asian Origin in Medina Gounass (Guediawaye), Senegal

Idrissa Dieng [1], Boris Gildas Hedible [2], Moussa Moïse Diagne [1], Ahmed Abd El Wahed [3], Cheikh Tidiane Diagne [1], Cheikh Fall [1], Vicent Richard [2], Muriel Vray [2], Manfred Weidmann [4], Ousmane Faye [1,*], Amadou Alpha Sall [1] and Oumar Faye [1,*]

[1] Arboviruses and Hemorrhagic Fever Viruses Unit, Virology Department, Institut Pasteur de Dakar, BP220 Dakar, Senegal; idrissa.dieng@pasteur.sn (I.D.); MoussaMoise.Diagne@pasteur.sn (M.M.D.); ct.Diagne@pasteur.sn (C.T.D.); Cheikh.Fall@pasteur.sn (C.F.); Amadou.SALL@pasteur.sn (A.A.S.)
[2] Epidemiology Unit, Institut Pasteur de Dakar, BP220 Dakar, Senegal; ghedible@pasteur.sn (B.G.H.); vrichard@pasteur.nc (V.R.); muriel.vray@pasteur.sn (M.V.)
[3] Microbiology and Animal Hygiene, University of Goettingen, D-33077 Goettingen, Germany; abdelwahed@dpz.eu
[4] Institute of Aquaculture, University of Stirling, Scotland FK9 4LA, UK; m.w.weidmann@stir.ac.uk
* Correspondence: Ousmane.FAYE@pasteur.sn (O.F.); Oumar.FAYE@pasteur.sn (O.F.)

Received: 21 May 2020; Accepted: 11 June 2020; Published: 16 June 2020

Abstract: With the growing success of controlling malaria in Sub-Saharan Africa, the incidence of fever due to malaria is in decline, whereas the proportion of patients with non-malaria febrile illness (NMFI) is increasing. Clinical diagnosis of NMFI is hampered by unspecific symptoms, but early diagnosis is a key factor for both better patient care and disease control. The aim of this study was to determine the arboviral aetiologies of NMFI in low resource settings, using a mobile laboratory based on recombinase polymerase amplification (RPA) assays. The panel of tests for this study was expanded to five arboviruses: dengue virus (DENV), zika virus (ZIKV), yellow fever virus (YFV), chikungunya virus (CHIKV), and rift valley fever virus (RVFV). One hundred and four children aged between one month and 115 months were enrolled and screened. Three of the 104 blood samples of children <10 years presented at an outpatient clinic tested positive for DENV. The results were confirmed by RT-PCR, partial sequencing, and non-structural protein 1 (NS1) antigen capture by ELISA (Biorad, France). Phylogenetic analysis of the derived DENV-1 sequences clustered them with sequences of DENV-1 isolated from Guangzhou, China, in 2014. In conclusion, this mobile setup proved reliable for the rapid identification of the causative agent of NMFI, with results consistent with those obtained in the reference laboratory's settings.

Keywords: fever; NMFI; mobile laboratory; RPA; DENV

---

1. Introduction

In Africa, fever is the most common symptom leading patients to seek health care [1,2]. Fever of unknown origin has long served as an entry point for the treatment of malaria [3]. With encouraging gains in malaria control in Sub-Saharan African countries, the incidence of this disease is in decline, leading to a decreasing proportion of febrile illness attributable to malaria. Between 2000 and 2013, malaria mortality rates decreased by 47% globally, and by 54% in sub-Saharan Africa—the region most affected by the disease—whereas the proportion of patients presenting with non-malaria febrile illness (NMFI) increased, respectively [4]. Acute febrile episodes are caused by various bacterial and viral pathogens, and infections with these agents result in patients presenting with malaria-like symptoms [5]. Although resulting in a higher mortality than malaria, NMFIs are not being reliably diagnosed due to the lack of accurate, rapid and affordable diagnostic tests, and also due to poor access to diagnostics facilities in many resource-poor endemic settings [1,6,7]. The objective of this study was to determine potential arboviral etiologies of NMFI in children in a low resource setting, using mobile recombinase polymerase amplification (RPA)—a real-time isothermal amplification technique [8]. For this purpose, we conducted a prospective arbovirus investigation in children seeking healthcare, at a health centre in the Dakar suburb of Medina Gounass, during a period of six months—from September 2015 to March 2016. The mobile suitcase laboratory has also been successfully used for Ebola virus detection [9].

2. Materials and Methods

2.1. Study Site

For a pilot study, prospective molecular screening on NMFI was conducted between September 2015 and March 2016, at the "Institut de pédiatrie sociale", located in the suburb of Dakar (Figure 1)—the capital city of Senegal, West Africa. Built in 1971, this health centre located at Pikine-Guediawaye has an outpatient department with care activities focused on mother and child health. It is also involved in the national program on immunization, in nutritional programs and in family planning. With 22 qualified staff, the health centre has 6 consultation rooms (including one for vaccination), a laboratory and a nutritional service. With around 1,000,000 inhabitants, Pikine-Guediawaye is an agglomeration of well-established traditional villages, and interspersed recent settlements, the latter mostly located in flood-prone areas, where housing is officially forbidden. The western part of these towns is located on the edge of a vast area of permanent marshland (Grande Niaye), where natural marshy hollows and furrows dug for market garden irrigation, as well as areas of prolonged stagnation of rainwater, are observed year-round. With a high density of housing (9200 inhabitants/km) in proximity to stretches of water and stagnant wetlands, the population lives in an under-serviced peripheral area, in crowded conditions, with poor water supply and sanitation, and dirt paths between dwellings and open sewers.

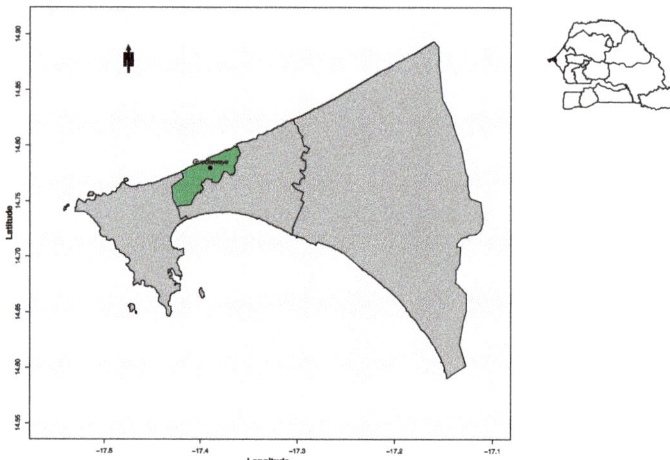

Figure 1. Map showing the area of study.

2.2. Patient Selection

Children less than 10 years old were enrolled if they met the following criteria: acute fever ($\geq$37.5 °C axillary temperature), negative for malaria rapid diagnostic tests (RDTs) and living in the same area for four successive calendar months. Regarding the eligibility for enrolment, the study information was read to the legal guardian, and after obtaining informed consent, clinical symptoms were recorded, and 2 mL of venal blood was collected. The study initially consisted of a weekly visit every Monday, before Friday was added to compensate for the small sampling observed at the beginning of the survey.

2.3. Screening Procedure in the Field

Blood samples were collected and processed on site, using a mobile suitcase laboratory for viral identification. The mobile laboratory consisted of a glovebox (Bodo Koennecke, Berlin, Germany), a lab-in-a-suitcase [9–11], a solar panel and a power pack set (Yeti 400 set, GOALZERO, South Bluffdale, UT, USA, Figure 2). The mobile setup was organized into 2 stations: the extraction station with the glovebox, and the RPA in the suitcase laboratory. Inactivation and extraction of the samples were performed in the glovebox using a modified protocol of the Speedxtract kit (Qiagen, Lake Constance, Germany).

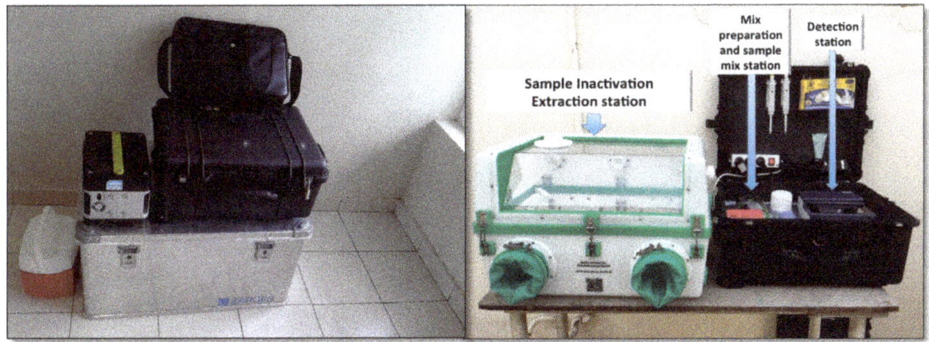

Figure 2. Deployment of the mobile laboratory at the Institut Pasteur de Dakar before departure (**left**), and at the site of study, Medina Gounass (**right**). For more detail on the extraction procedure, see Figure 2 in [11].

The extraction was performed by adding 100 µL of lysis buffer, 20 µL of sera and 30 µL of magnetic beads to a 1.5 mL tube, followed by incubation at 95 °C for 10 min. After incubation, the tube was transferred to a magnetic stand for 2 min, and after sedimentation of the beads, a 150 µL volume of supernatant was collected in a new 1.5 mL tube. RNA amplification of all samples was carried out using the Tubescanner point of care device (Qiagen, Hilden, Germany), and the Twist exo RT kit (TwistDx, Cambridge, UK), using RPA amplicons designed for the detection of 5 arboviruses: dengue virus (DENV), with two sets of primers and probes (set 1 for DENV1-3 and set 2 for DENV- 4 [10]); yellow fever virus (YFV) [12]; chikungunya virus (CHIKV) [13]; rift valley fever virus (RVFV) [14]; and zika virus (ZIKV) [15], with analytical sensitivities of 241, 14, 10, 23, 10, 21; and RNA molecules were detected, respectively. RT-RPA reactions were performed in a volume of 50 µL. Briefly, a mix containing 29.5 µL of rehydration buffer, 7.2 µL of ddH$_2$O, 420 nM of each primer, and 210 nM of a target specific RPA exo-probe, was dispensed into each of the eight 1.5 mL tubes, before adding 5 µL of RNA template. Finally, 46.5 µL of master mix/template solution was transferred to each lyophilized RPA pellet of the eight-tubes strip provided in the kit. An amount of 3.5 µL magnesium acetate (280 mM) was added into the lid of each tube, before closing it and spinning the drop into the reaction mix. For real-time fluorescence monitoring, the reaction tubes were placed in the ESE Quant Tubescanner (Qiagen Lake Con- stance GmbH, Stockach, Germany).

2.4. Standard Assays in the Central Laboratory

All of the samples collected in the field were shipped to the WHO collaborating centre for arboviruses and haemorrhagic fever viruses, at the Institut Pasteur de Dakar (IPD), in order to perform complementary confirmatory tests.

2.4.1. RNA Extraction and Real-Time qPCR

Viral RNA was extracted from 100 μL of human sera, using the QIAmp viral RNA kit (Qiagen, Hilden, Germany) according to the manufacturer's instructions. The reduced input of 100 μL serum for this kit had been validated in-house. The RNA was eluted in 60 μL of elution buffer and placed on ice. To confirm the detection obtained in the field, real-time RT-qPCR assays for DENV [16], CHIKV [17], RVFV [18], YFV [19], ZIKV [20] were performed using the Quantitect Probe RT-PCR Master Mix (Qiagen). Briefly, the detection was performed using ABI7500, using the following temperature profiles for all RT-qPCR assays: RT at 50 °C for 10 min, activation at 95 °C for 15 min and 45 cycles of 2-step PCR— at 95 °C for 15 sec and 60 °C for 1 min.

2.4.2. Non-Structural Protein 1 (NS1) Antigen Capture

The NS1 antigen assay was performed using the Platelia™ Dengue NS1 Ag-ELISA kit (Biorad Laboratories, Marnes-La-Coquette, France), according to the manufacturer's instructions. The capture and revelation steps were performed using a murine monoclonal antibody (MAb). The presence of the NS1 antigen in a sample was assessed by the formation of an immune-complex MAb-NS1-MAb (peroxidase). Briefly, 50 μL of the serum sample, 50 μL of the sample diluent (Diluent R7), and 100 μL of the diluted conjugate were added to each well precoated with the anti-NS1 monoclonal antibody. For each assay, positive and negative controls, as well as calibrator sera, were included. The plate was incubated for 90 min at 37 °C. After six wash steps—using 250 μL of washing solution (R2)—160 μL of tetramethylbenzidine (TMB) substrate was added to each of the wells, and the plate was further incubated at room temperature for 30 min in the dark, followed by the addition of 100 μL of stop solution (1N H2SO4). With the spectrophotometer optical density (OD) readings were measured at wavelengths of 450 nm/620 nm, were obtained and the index of each sample was calculated with the ratio OD of samples/OD of calibrators. Sample ratios <0.5, between 0.5 and 1.0, and >1.0 were classified as negative, equivocal, and positive, respectively.

2.4.3. ELISA IgM Detection

We determined the presence of DENV, YFV, RVFV, CHIKV and ZIKV IgM in our samples by a capture enzyme-linked immunosorbent assay (MAC ELISA), following a published protocol [21]. For the coating step, a monoclonal capture antibody (anti human IgM) was added to a 96-well microtiter plate, and incubated at 4 °C overnight. The human sera were heat-inactivated (56 °C, 30 min), and screened at a dilution of 1:100 in phosphate-buffered saline (PBS), supplemented with 0.05% Tween and 1% milk powder. After washing the plate three times with PBS plus 0.05% Tween 20, a 1/100 dilution of the different serum samples and controls were added in duplicate to the plate and incubated at 37 °C for 1 h. The wells were washed three times, specific and non-specific antigens were deposited, and the plate was incubated for 1 h at 37 °C. After another washing step, the specific immune ascites were added to each well. After incubation and washing, a conjugate (peroxidase labelled antibody specific to mouse IgG) was added and allowed to react for 1 h at 37 °C. A tetra-methylbenzidine (TMB) substrate was added to the IgM conjugate complex, and the coloration reaction was stopped with sulphuric acid. The intensity of the coloration was proportional to the level of virus specific antibodies present in the serum. An ELISA microplate reader showed the optical density (OD) of the absorbance—an OD unit ≥ 0.2 was defined as a positive IgM.

2.4.4. Viral Isolation and Immunofluorescence Indirect Assay

Virus isolation was attempted for samples positive with RPA/RT-qPCR. A 200 µL sample was diluted at 1:10 in a Leibovitz 15 medium (L15), then added to a 25 cm^2 flask monolayer of C6/36 cell line at 80% confluence, followed by incubation at 28 °C for 1 h, to allow virus adsorption. After incubation, the L15 medium containing 5% FBS, 1% penicillin streptomycin, and 0.05% de fungizone, was added to the flask and incubated for 10 days, or until observation of a cytopathic effect (CPE). To assess viral infection, indirect immunofluorescence (IFA) was conducted as previously described [22]. The flask content was transferred to a 15 mL tube, and clarified by low-speed (2500 rpm) centrifugation at 4 °C for 5 min. The supernatant was harvested and stored at −80 °C, while the cells were washed three times in PBS 1×, resuspended in 4 mL of PBS 1×, and then dispensed onto a glass slide. After complete drying, the cells were fixed in cold (−20°) acetone for 20 min. Staining was performed with a DENV specific hyper-immune mouse ascitic fluid, diluted at 1/40 in PBS 1×, as the first antibody. Then, the cells were incubated for 30 min with the second antibody (1/40 goat anti-mouse IgG, 1/100 Evan's blue, diluted in PBS 1×). The slides were observed with a fluorescence microscope (Eurostar III plus, Euroimmune).

2.4.5. RT-PCR Amplification, Sequencing and Phylogenetic Analysis

To define the serotype of the isolated DENV strains, viral RNA was extracted from 200 µL of DENV infected C6/36 culture supernatant, using the QIAmp viral RNA kit (Qiagen, Hilden, Germany) in accordance with the manufacturer's instructions. The C-prM gene was amplified using the primers (DS1/DS2) described by Lanciotti et al. [23]. For cDNA synthesis, 10 µL of viral RNA was mixed with 1 µL of the random hexamer primer (2 pmol), and the mixture was heated at 95 °C for 2 min. The reverse transcription was performed in a 20 µL mixture containing 2.5 U RNasin (Promega, Madison, WI, USA), 1 µL of deoxynucleotide triphosphate (dNTP) (10 mM each DNTP), and 5 U of AMV reverse transcriptase (Promega, Madison, WI, USA), which was incubated at 42 °C for 60 min. The PCR products were generated using DS1/DS2 primers at 10 µM concentration. Five microliters of cDNA were mixed with 10× buffer, 3 µL of each primer, 5 µL of dNTPs 10 mM, 3 µL of MgCl2, and 0.5 µL of Taq polymerase (Promega, Madison, WI, USA). Amplicons of the expected size (520 bp) were purified from the agarose gel, with the QiaQuick Gel Extraction Kit (Qiagen, Hilden, Germany), as specified by the manufacturer. Both strands of each amplicon were Sanger sequenced out-of-house (Genewiz, Germany). The sequences were merged using EMBOSS Merger software, and final results were analyzed using the Basic Local Alignment Search Tool (BLAST, www.ncbi.nlm.nih.gov/) consulted on 18 July 2017. The nucleotide sequence alignments were generated using the ClustalW algorithm, and Maximum likelihood (ML) trees were inferred for each serotype using Mega software version 6 [24].

3. Results

3.1. Demographic and Clinical Data

During the study period from September 2015 to March 2016, 104 children aged between 1 and 115 months were enrolled and screened for DENV, CHIKV, ZIKV, YFV, RVFV—Seventy-nine were <5 years old, and 42 (40,38%) were male (Supplementary Data 1). Regarding clinical symptoms, the most common symptoms after fever were, respectively, rhinorrhea (65%), cough (53%; 55/104), vomiting (25%; 26/104), diarrhoea (25%; 25/104) and abdominal pain (25%; 26/104). Headaches were reported only in 10% of the enrolled patients, while myalgia and arthralgia were not recorded at all (Figure 3).

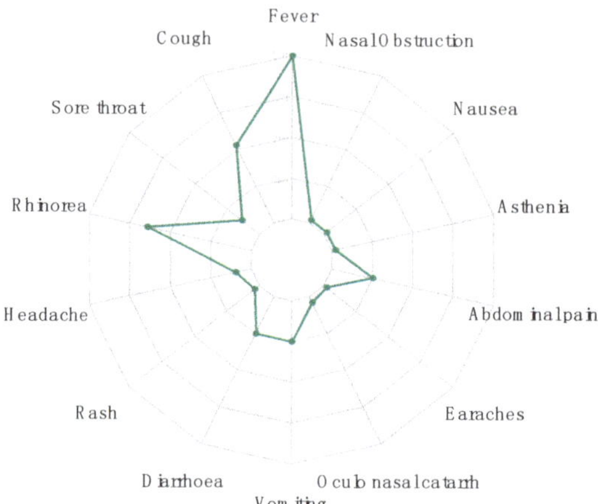

Figure 3. Plot showing the percentage of recorded symptoms among enrolled patients.

3.2. Molecular, Serologic and Antigenic Detection

Three samples (2.8%) tested positive with DENV1-3 RPA, while none of the other arbovirus assays yielded positive results. All of the arbovirus results were confirmed by RT-PCR, with 100% concordance. The samples of the three DENV cases were additionally tested and confirmed positive by DENV NS1 antigen capture (Table 1). The ELISA IgM test yielded negative results in all three cases (not shown). Virus isolation at 28 °C in C6/36 cells yielded non-obvious CPE ten days after inoculation. However, two strains were successfully isolated, and isolation was confirmed by IFA (Figure S1). The isolates were designated Medina Gounass 1 and Medina Gounass 2, respectively.

Table 1. Comparison of time detection between non-structural protein 1 (NS1), real-time recombinase polymerase amplification (RT-RPA), and RT-PCR.

Sample Names	RT-qPCR		RT-RPA		NS1 Antigen Capture	
	Ct Value	Detection Time	Threshold Time	Detection Time	D.O	Average Detection Time
Medina Gounass 1	29.35	79 min	6 min	6 min	6.8477	2 h
Medina Gounass 2	27.59	75 min	5 min	5 min	7.9131	2 h
Medina Gounass 3	26.42	72 min	5 min	5 min	9.819	2 h

3.3. Phylogenetic Analysis

Finally, the sequence of the CprM gene of the two isolates was determined and deposited in genbank (accessions numbers MK940790 (Medina gounass 1/2015), and MK940791 (Medina gounass 1/2015)). Both strains were completely homologous (100%), with no nucleotide difference. A basic local alignment search tool for nucleotide (BLASTN), using the obtained C-prM gene sequence, showed 100% identity with the DENV-1 isolates collected in Guangzhou, China, in 2014 (KP279753.1). A calculated phylogenetic tree clustered the determined DENV-1 sequences with Asian strains, supported by high bootstrap values (Figure 4).

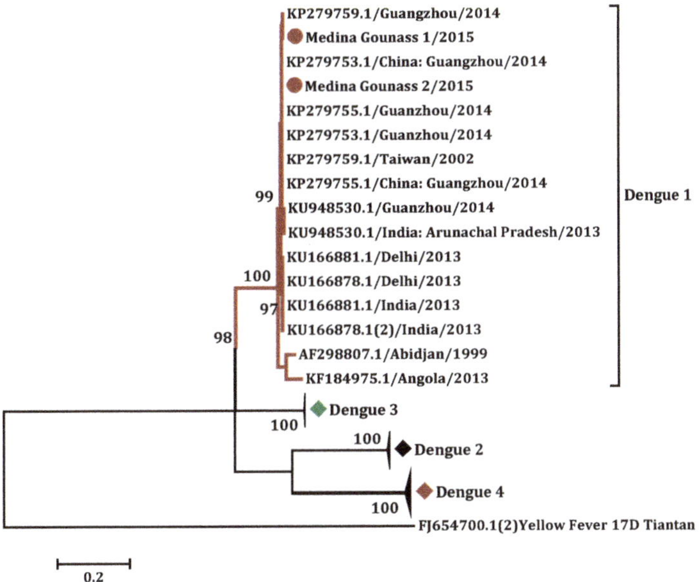

Figure 4. Phylogenetic tree based on the C-prM gene sequences using the maximum likelihood (ML) method, showing the relationship of 2 from 3 isolated viruses in this study (darkred circle) with 40 global samples of dengue virus (DENV), and Kimura 2 parameters; the gamma distributed (K2+G+I) nucleotide substitution model was used. The yellow fever 17D (FJ654700.1) was used as the outgroup.

4. Discussion

The objective of this study was to assess the use of a mobile suitcase laboratory for the routine testing of arboviral etiologies of NMFI in an outpatient clinic of a suburb of Dakar, Senegal. During the six month survey of arboviral infections in febrile non malaria patients, three cases of dengue infection were detected in 104 enrolled children under 10 years old.

None of the targeted arboviruses (CHIKV, RVFV, YFV, ZIKV) except for DENV were detected during this study. The prevalence of DENV was 2.8%.

In a previous study on the etiology of acute febrile illness in Abidjan, an inter epidemic DENV prevalence of 0.4% was reported in the 812 patients tested [25]. Similarly, our work highlights an inter epidemic circulation of DENV in poor urban settings of Dakar. The difference in prevalence between the two studies may be attributable to the fact that the study conducted in Abidjan was not limited to children ≤10 years of age, as well as the smaller number of enrolled patients in our study (104 patients).

The border between interepidemic and epidemic prevalence in sub-Saharan Africa is difficult to assess, as noted by a study on febrile patients in Ibadan, Nigeria [26], which determined a prevalence of 35% of dengue infection through NS1 antigen detection. The infected patients secrete large quantities of soluble NS1 (sNS1) into the bloodstream, with concentrations of up to 50 µg/mL [27]. Soluble NS1 (sNS1) can remain in the blood for 9 days, and persist for up to 18 days in some patients [28], exceeding viremia which lasts up to 6 days. This makes NS1 a good biomarker of acute illness as it provides a wide window for DENV detection. It has been suggested that combining NS1 detection with IgM detection can outperform PCR [29]; however, the use of NS1 detection in the routine screening in dengue epidemics, as a prerequisite for hospitalization, has been questioned [30]. Additionally, fieldable ELISA systems which would allow for a comparison between the ratios of DENV NS1-Ag and DENV-RNA, are not currently available.

Phylogenetic analysis of the obtained DENV C-prM gene sequences yielded 100% identity with the isolates collected in Guangzhou, China, in 2014. A calculated phylogenetic tree clustered the determined DENV-1 sequences with Asian strains, supported by high bootstrap values (Figure 4). This finding suggests an importation of the virus to Senegal from Asia, via acutely viremic cases or by infected mosquitoes or their eggs. Indeed, in recent years, international travel and trade activity between the Asian and African continents has increased considerably [31]—between 1994 and 2009, the annual volume of trade between Senegal and China grew from 23 million U.S. dollars to 441 million U.S. dollars, representing a twenty-fold increase in 15 years [32]. The potential to extend the distribution area of individual arboviruses was recently supported by the detection of Japanese encephalitis virus (JEV), during a yellow fever outbreak in Lunda (Angola), in 2016 [33]. This virus is endemic in Asia and the western pacific, but local circulation had never been documented before in Africa [34]. Another example is the first case of RVFV infection, detected in China from a patient returning from Angola in 2017—while this virus was previously restricted to sub-Saharan Africa, it has been spreading in the Arabic peninsula since 2000 [35].

In Africa, dengue is likely to be underreported and under-recognized. This is due to the low awareness of health care providers, and the circulation of other prevalent NMFI [5]. The absence of surveillance in many African countries and the lack of effective diagnostic tools also contribute to the underestimation of the real incidence of dengue fever in Africa [31]. Since other studies report the expansion of dengue fever among NMFI [36], our study and the cited studies in Abidjan and Ibadan stress the need to implement laboratory capacity to assess the real burden of DENV in rural and urban areas of West Africa, during inter epidemic periods.

The RPA positive samples were confirmed by serological assay, viral isolation as well as real-time RT-PCR. The laboratory-based real-time RT-PCR and mobile RT-RPA results were concordant, but mobile RT-RPA yielded results in approximately 20 min, including the extraction step (Table 1). Additionally, the RPA was performed at the point of need in a suitcase laboratory. In conclusion, although virus isolation remains the "gold standard" in diagnostics [37], rapid molecular testing at the point of care can provide reliable results (short time-length process, sensitivity and specificity).

5. Conclusions

Our results suggest that the RT-RPA could be an alternative to real-time PCR in low resource settings. This field deployment contributed to the evaluation of the feasibility to implement point of need arbovirus diagnostics in primary care settings and showed that RT-RPA can be a reliable and accurate diagnosis tool for the detection of NMFI in low-income settings. However, studies with larger cohorts are needed.

Supplementary Materials: The following are available online at http://www.mdpi.com/2075-4418/10/6/408/s1, Figure S1: Assessment of Dengue virus infection using immunofluorescence Indirect Assay (IFA) on C6/36 cells infected with 200 µL of crushed mouse brain previously inoculated with Positive Sera. Immune Ascite fluid was used as the Primary antibody (A: Negative control, B: Medina Gounass 1, C: Medina Gounass 2), Supplementary Data 1: Data table supporting the findings.

Author Contributions: Conceptualization, A.A.S., V.R. and O.F. (Oumar Faye); data curation, I.D., O.F. (Oumar Faye), A.A.E.W. and B.G.H.; formal analysis, I.D., M.M.D., and O.F. (Oumar Faye); funding acquisition, V.R. and A.A.S.; investigation, I.D., B.G.H. and M.M.D.; methodology, I.D., B.G.H., M.W., A.A.E.W., O.F. (Oumar Faye) and A.A.S.; project administration, A.A.S.; resources, O.F. (Oumar Faye), V.R. and A.A.S.; software, I.D., supervision, O.F. (Oumar Faye), M.W., M.V., O.F. (Ousmane Faye), and A.A.S.; visualization, I.D.; writing—original draft, I.D.; writing—review and editing, I.D., M.M.D., B.G.H., O.F. (Ousmane Faye), O.F. (Oumar Faye), C.T.D., C.F., and M.W. All authors have read and agreed to the published version of the manuscript.

Funding: This work was supported by a grant from Clayton Dedonder (EC/MAM/N° 323/14) and by the Institut Pasteur de Dakar funds.

Acknowledgments: We thank all of the workers in the virology department of IPD (Institut Pasteur de Dakar), as well as the staff of the Medina Gounass Health Center.

Conflicts of Interest: The authors declare no conflict of interests.

References

1. Petti, C.A.; Polage, C.R.; Quinn, T.C.; Ronald, A.R.; Sande, M.A. Laboratory Medicine in Africa: A Barrier to Effective Health Care. *Clin. Infect. Dis.* **2006**, *42*, 377–382. [CrossRef] [PubMed]
2. Feikin, D.R.; Olack, B.; Bigogo, G.M.; Audi, A.; Cosmas, L.; Aura, B.; Burke, H.; Njenga, M.K.; Williamson, J.; Breiman, R.F. The Burden of Common Infectious Disease Syndromes at the Clinic and Household Level from Population-Based Surveillance in Rural and Urban Kenya. *PLoS ONE* **2011**, *6*, e16085. [CrossRef] [PubMed]
3. Okiro, E.A.; Snow, R.W. The Relationship between reported fever and Plasmodium falciparum infection in African children. *Malar J.* **2010**, *9*, 99. [CrossRef]
4. WHO. *World Malaria Report 2014*; World Health Organization: Geneva, Switzerland, 2014.
5. Were, T.; Davenport, G.C.; Hittner, J.B.; Ouma, C.; Vulule, J.M.; Ong'Echa, J.M.; Perkins, D.J. Bacteremia in Kenyan Children Presenting with Malaria. *J. Clin. Microbiol.* **2010**, *49*, 671–676. [CrossRef] [PubMed]
6. Tadesse, H. The Etiology of Febrile Illnesses among Febrile Patients Attending Felegeselam Health Center, Northwest Ethiopia. *Am. J. Biomed. Life Sci.* **2013**, *1*, 58. [CrossRef]
7. Crump, J.A.; Ramadhani, H.O.; Morrissey, A.B.; Saganda, W.; Mwako, M.S.; Yang, L.-Y.; Chow, S.-C.; Morpeth, S.C.; Reyburn, H.; Njau, B.N.; et al. Invasive Bacterial and Fungal Infections Among Hospitalized HIV-Infected and HIV-Uninfected Adults and Adolescents in Northern Tanzania. *Clin. Infect. Dis.* **2011**, *52*, 341–348. [CrossRef]
8. Piepenburg, O.; Williams, C.H.; Stemple, D.L.; Armes, N.A. DNA Detection Using Recombination Proteins. *PLoS Boil.* **2006**, *4*, e204. [CrossRef]
9. Faye, O.; Faye, O.; Soropogui, B.; Patel, P.; El Wahed, A.A.; Loucoubar, C.; Fall, G.; Kiory, D.; Magassouba, N.; Keita, S.; et al. Development and deployment of a rapid recombinase polymerase amplification Ebola virus detection assay in Guinea in 2015. *Eurosurveillance* **2015**, *20*. [CrossRef]
10. El Wahed, A.A.; Patel, P.; Faye, O.; Thaloengsok, S.; Heidenreich, D.; Matangkasombut, P.; Manopwisedjaroen, K.; Sakuntabhai, A.; Sall, A.A.; Hufert, F.T.; et al. Recombinase Polymerase Amplification Assay for Rapid Diagnostics of Dengue Infection. *PLoS ONE* **2015**, *10*, e0129682. [CrossRef]
11. Weidmann, M.; Faye, O.; Faye, O.; El Wahed, A.A.; Patel, P.; Batejat, C.; Manugerra, J.C.; Adjami, A.; Niedrig, M.; Hufert, F.T.; et al. Development of Mobile Laboratory for Viral Hemorrhagic Fever Detection in Africa. *J. Infect. Dis.* **2018**, *218*, 1622–1630. [CrossRef]
12. Escadafal, C.; Faye, O.; Sall, A.A.; Faye, O.; Weidmann, M.; Strohmeier, O.; Von Stetten, F.; Drexler, J.; Eberhard, M.; Niedrig, M.; et al. Rapid Molecular Assays for the Detection of Yellow Fever Virus in Low-Resource Settings. *PLoS Negl. Trop. Dis.* **2014**, *8*, e2730. [CrossRef] [PubMed]
13. Patel, P.; El Wahed, A.A.; Faye, O.; Prüger, P.; Kaiser, M.; Thaloengsok, S.; Ubol, S.; Sakuntabhai, A.; Leparc-Goffart, I.; Hufert, F.T.; et al. A Field-Deployable Reverse Transcription Recombinase Polymerase Amplification Assay for Rapid Detection of the Chikungunya Virus. *PLoS Negl. Trop. Dis.* **2016**, *10*, e0004953. [CrossRef] [PubMed]
14. Euler, M.; Wang, Y.; Nentwich, O.; Piepenburg, O.; Hufert, F.T.; Weidmann, M. Recombinase polymerase amplification assay for rapid detection of Rift Valley fever virus. *J. Clin. Virol.* **2012**, *54*, 308–312. [CrossRef] [PubMed]
15. El Wahed, A.A.; Sanabani, S.S.; Faye, O.; Pessôa, R.; Patriota, J.V.; Giorgi, R.R.; Patel, P.; Böhlken-Fascher, S.; Landt, O.; Niedrig, M.; et al. Rapid Molecular Detection of Zika Virus in Acute-Phase Urine Samples Using the Recombinase Polymerase Amplification Assay. *PLoS Curr.* **2017**, *9*. [CrossRef]
16. Wagner, D.; De With, K.; Huzly, D.; Hufert, F.; Weidmann, M.; Breisinger, S.; Eppinger, S.; Kern, W.V.; Bauer, T.M. Nosocomial Acquisition of Dengue. *Emerg. Infect. Dis.* **2004**, *10*, 1872–1873. [CrossRef]
17. Shu, P.-Y.; Yang, C.-F.; Su, C.-L.; Chen, C.-Y.; Chang, S.-F.; Tsai, K.-H.; Cheng, C.-H.; Huang, J.-H. Two Imported Chikungunya Cases, Taiwan. *Emerg. Infect. Dis.* **2008**, *14*, 1325–1326. [CrossRef]
18. Weidmann, M.; Sanchez-Seco, M.P.; Sall, A.A.; Ly, P.O.; Thiongane, Y.; Lô, M.M.; Schley, H.; Hufert, F.T. Rapid detection of important human pathogenic Phleboviruses. *J. Clin. Virol.* **2008**, *41*, 138–142. [CrossRef]
19. Weidmann, M.; Faye, O.; Faye, O.; Kranaster, R.; Marx, A.; Nunes, M.R.T.; Vasconcelos, P.F.; Hufert, F.T.; Sall, A.A. Improved LNA probe-based assay for the detection of African and South American yellow fever virus strains. *J. Clin. Virol.* **2010**, *48*, 187–192. [CrossRef]
20. Faye, O.; Faye, O.; Diallo, D.; Diallo, D.; Weidmann, M.; Sall, A.A. Quantitative real-time PCR detection of Zika virus and evaluation with field-caught Mosquitoes. *Virol. J.* **2013**, *10*, 311. [CrossRef]

21. Ba, F.; Loucoubar, C.; Faye, O.; Fall, G.; Mbaye, R.N.P.N.; Sembene, M.; Diallo, M.; Balde, A.T.; Sall, A.A.; Faye, O. Retrospective analysis of febrile patients reveals unnoticed epidemic of zika fever in Dielmo, Senegal, 2000. *Clin. Microbiol. Infect. Dis.* **2018**, *3*. [CrossRef]

22. Digoutte, J.; Calvo-Wilson, M.; Mondo, M.; Traore-Lamizana, M.; Adam, F. Continuous cell lines and immune ascitic fluid pools in arbovirus detection. *Res. Virol.* **1992**, *143*, 417–422. [CrossRef]

23. Lanciotti, R.S.; Calisher, C.H.; Gubler, D.J.; Chang, G.J.; Vorndam, A.V. Rapid detection and typing of dengue viruses from clinical samples by using reverse transcriptase-polymerase chain reaction. *J. Clin. Microbiol.* **1992**, *30*, 545–551. [CrossRef] [PubMed]

24. Tamura, K.; Stecher, G.; Peterson, D.; Filipski, A.; Kumar, S. MEGA6: Molecular Evolutionary Genetics Analysis Version 6.0. *Mol Biol Evol.* **2013**, *30*, 2725–2729. [CrossRef] [PubMed]

25. L'Azou, M.; Succo, T.; Kamagaté, M.; Ouattara, A.; Gilbernair, E.; Adjogoua, E.; Luxemburger, C. Dengue: Etiology of acute febrile illness in Abidjan, Côte d'Ivoire, in 2011-2012. *Trans. R. Soc. Trop. Med. Hyg.* **2015**, *109*, 717–722. [CrossRef] [PubMed]

26. Oyero, O.; Ayukekbong, J.A. High dengue NS1 antigenemia in febrile patients in Ibadan, Nigeria. *Virus Res.* **2014**, *191*, 59–61. [CrossRef]

27. Libraty, D.H.; Young, P.R.; Pickering, D.; Endy, T.P.; Kalayanarooj, S.; Green, S.; Vaughn, D.W.; Nisalak, A.; Ennis, F.A.; Rothman, A.L. High Circulating Levels of the Dengue Virus Nonstructural Protein NS1 Early in Dengue Illness Correlate with the Development of Dengue Hemorrhagic Fever. *J. Infect. Dis.* **2002**, *186*, 1165–1168. [CrossRef] [PubMed]

28. Amorim, J.H.; Alves, R.P.D.S.; Boscardin, S.B.; Ferreira, L.C.D.S. The dengue virus non-structural 1 protein: Risks and benefits. *Virus Res.* **2014**, *181*, 53–60. [CrossRef] [PubMed]

29. Anand, A.M.; Sistla, S.; Dhodapkar, R.; Hamide, A.; Biswal, N.; Srinivasan, B. Evaluation of NS1 Antigen Detection for Early Diagnosis of Dengue in a Tertiary Hospital in Southern India. *J. Clin. Diagn. Res.* **2016**, *10*, DC01–DC04. [CrossRef] [PubMed]

30. Pothapregrada, S. The Dilemma of Reactive NS1 Antigen Test in Dengue Fever. *Indian Pediatr.* **2015**, *52*, 906–907.

31. Amarasinghe, A.; Kuritsky, J.N.; Letson, G.W.; Margolis, H.S. Dengue Virus Infection in Africa. *Emerg. Infect. Dis.* **2011**, *17*, 1349–1354. [CrossRef]

32. Gehrold, S. Far from altruistic: China's presence in Senegal. 29. Available online: https://www.kas.de/c/document_library/get_file?uuid=a2cb2655-3c82-ad66-4668-8300fdac0bff&groupId=252038 (accessed on 10 June 2020).

33. Simon-Loriere, E.; Kipela, J.-M.; Fernandez-Garcia, M.D.; Diagne, M.M.; Fall, I.S.; Holmes, E.; Sall, A.A.; Faye, O.; Prot, M.; Casadémont, I.; et al. Autochthonous Japanese Encephalitis with Yellow Fever Coinfection in Africa. *N. Engl. J. Med.* **2017**, *376*, 1483–1485. [CrossRef] [PubMed]

34. Mansfield, K.; Hernández-Triana, L.; Banyard, A. Japanese encephalitis virus infection, diagnosis and control in domestic animals|Elsevier Enhanced Reader. *Vet. Microbiol.* **2017**, *201*, 85–92. [CrossRef] [PubMed]

35. Liu, J.; Sun, Y.; Shi, W.; Tan, S.; Pan, Y.; Cui, S.; Zhang, Q.; Dou, X.; Lv, Y.; Li, X.; et al. The first imported case of Rift Valley fever in China reveals a genetic reassortment of different viral lineages. *Emerg. Microbes Infect.* **2017**, *6*, e4–e7. [CrossRef] [PubMed]

36. Stanaway, J.D.; Shepard, N.S.; Undurraga, E.A.; Halasa, Y.A.; Coffeng, L.E.; Brady, O.J.; Hay, S.I.; Bedi, N.; Bensenor, I.M.; Castañeda-Orjuela, C.; et al. The global burden of dengue: An analysis from the Global Burden of Disease Study 2013. *Lancet Infect. Dis.* **2016**, *16*, 712–723. [CrossRef]

37. Leland, D.S.; Ginocchio, C.C. Role of Cell Culture for Virus Detection in the Age of Technology. *Clin. Microbiol. Rev.* **2007**, *20*, 49–78. [CrossRef] [PubMed]

Article

Development of a Flow-Free Automated Colorimetric Detection Assay Integrated with Smartphone for Zika NS1

Md Alamgir Kabir [1,2], Hussein Zilouchian [2], Mazhar Sher [1,2] and Waseem Asghar [1,2,3,*]

[1]   Department of Computer & Electrical Engineering and Computer Science, Florida Atlantic University, Boca Raton, FL 33431, USA; mkabir2016@fau.edu (M.A.K.); msher2015@fau.edu (M.S.)
[2]   Asghar-Lab, Micro and Nanotechnology in Medicine, College of Engineering and Computer Science, Boca Raton, FL 33431, USA; hzilouchi@gmail.com
[3]   Department of Biological Sciences (Courtesy Appointment), Florida Atlantic University, Boca Raton, FL 33431, USA
*   Correspondence: wasghar@fau.edu

Received: 10 December 2019; Accepted: 12 January 2020; Published: 14 January 2020

Abstract: The Zika virus (ZIKV) is an emerging flavivirus transmitted to humans by *Aedes* mosquitoes that can potentially cause microcephaly, Guillain–Barré Syndrome, and other birth defects. Effective vaccines for Zika have not yet been developed. There is a necessity to establish an easily deployable, high-throughput, low-cost, and disposable point-of-care (POC) diagnostic platform for ZIKV infections. We report here an automated magnetic actuation platform suitable for a POC microfluidic sandwich enzyme-linked immunosorbent assay (ELISA) using antibody-coated superparamagnetic beads. The smartphone integrated immunoassay is developed for colorimetric detection of ZIKV nonstructural protein 1 (NS1) antigen using disposable chips to accommodate the reactions inside the chip in microliter volumes. An in-house-built magnetic actuator platform automatically moves the magnetic beads through different aqueous phases. The assay requires a total of 9 min to automatically control the post-capture washing, horseradish peroxidase (HRP) conjugated secondary antibody probing, washing again, and, finally, color development. By measuring the saturation intensity of the developed color from the smartphone captured video, the presented assay provides high sensitivity with a detection limit of 62.5 ng/mL in whole plasma. These results advocate a great promise that the platform would be useful for the POC diagnosis of Zika virus infection in patients and can be used in resource-limited settings.

Keywords: Zika NS1; point of care; colorimetric; smartphone; microfluidic

---

1. Introduction

The Zika virus (ZIKV) is a flavivirus that is closely related to other flaviviruses such as Dengue, West Nile, and Japanese encephalitis [1]. The virus is primarily spread through a vector, an infected *Aedes* species mosquito, but it can also be transferred through sexual contact and from a pregnant woman to their offspring [2]. Although there were only 1465 reported cases of the Zika virus in the continental USA, Central America, and Mexico, ZIKV still affects 87 countries as of September 2019, according to the World Health Organization (WHO) [3,4]. The ZIKV outbreak in the Americas occurred in 2016 with a steep decline in outbreaks in the following years, where in 2018 only 31,587 reported/probable cases were seen [4]. However, ZIKV is much more prevalent in some Asian and African countries. Indonesia has shown that 9.1% of its population under the age of five has had a prior ZIKV infection. In Lao People's Republic, 10% of the adult blood donor population showed a prior ZIKV infection [4]. Some symptoms of the ZIKV include mild fever, rash, conjunctivitis, and joint or muscle pain. Many individuals who are infected with ZIKV show no to mild symptoms, which can lead many patients to believe that they are not infected with the virus [5].

The lack of detection of ZIKV in an area cannot be necessarily equated to low levels of transmission or low levels of prevalence. The means of testing in certain areas, especially in rural and third world countries can be lacking, which can attribute to some of these findings. The nonstructural protein 1 (NS1) is targeted for detection due to its important role as a biomarker in flaviviruses. The protein is secreted by cells that contain the virus [6]. As the primary antigen, antibodies to NS1 can be formed in four to seven days, which can be used for detection. Currently the Centers for Disease Control and Prevention (CDC) and WHO guidelines to detect ZIKV are through the use of the nucleic acid amplification test (NAAT) [7]. RT-PCR is presently the most used NAAT within seven days of the onset of symptoms [8]. The problem NAAT faces, in general, is that it is lab based and cannot be performed at point-of-care (POC) settings, it also requires expensive reagents/equipment and technical staff. A negative NAAT) should also be cautiously considered. The IgM antibodies are usually detectable within four to seven days, so the WHO recommends serology testing seven days after the onset of symptoms. A standard enzyme-linked immunosorbent assay (ELISA) test requires 12 hours of testing and the expense of the reagents adds up over time. Therefore, reliable and inexpensive POC testing is of the utmost importance.

IgG and IgM antibodies are produced in the human body at the later stage of the infection (four to seven days), this makes them inapt for early stage detection. In contrast, NS1 antigen with a similar structure has been considered for highly sensitive, specific early stage detection for different flaviviruses previously [9–14]. There are currently numerous POC devices that detect ZIKV by using NS1 as a biomarker. The systematic evolution of ligands by exponential enrichment (SELEX) protocol uses aptamers to replace antibodies [10]. The aptamers used allowed the ZIKV NS1 antibody to bind with them, which could be detected through the use of ELISA. Although the technique lowered the cost and sensitivity, the assay is not suitable for point-of-care applications. Paper-based lateral flow immunoassay (LFIA) devices also exist, which can be effective in detecting infections/molecules [11,15–18]. Even though LFIA is cheap, rapid, and portable, it can only give qualitative or semiqualitative results. A laser cut glass fiber paper-based analytical device called PAD based on lateral flow technique has shown a limit of detection of around 25 ng/mL and can be used for NS1 biomarker detection in low-resource areas. The PAD assay provides only qualitative results, which can be done in 10 min [19]. However, the PAD requires several manual steps as well as a heating (60 °C) step to perform the assay.

In this paper, we have developed a microfluidic magnetic ELISA (M-ELISA) system that provides benefits over other POC testing and traditional testing through its low cost, time efficiency, and automation. Compared to conventional sandwich ELISA, the unique shape of the microfluidic chip that we have developed can decrease cost due to the use of less reagents and time by cutting the test from 12 hours to approximately 10 min. This can be beneficial in certain parts of the world where accessibility and cost are significant factors. The reduction in time is achieved through the increase of surface area with the magnetic beads [20]. The developed device is also automated, which can create a significant advantage and allow an individual to run multiple assays at a time.

2. Experimental Section

2.1. Sandwich ELISA for Detecting Zika NS1 Antigen in 96-Well-Plate Format

To validate the Antigen- Antibody (Ag–Ab) reaction, sandwich ELISA assay was first performed by coating a microplate with 100 µL of 2 µg/mL capture antibody for Zika NS1 (BF-1225-36, Biofront Technologies, Tallahassee, Florida, USA) in carbonate/bicarbonate buffer (pH 9.6), and then incubating it overnight at 4 °C. After washing three times with phosphate buffered saline (PBS) (pH 7.4), each well was blocked by 200 µL of SuperBlock T20 (PBS) Blocking Buffer (37517, Fisher Scientific, Hampton, NH, USA) for 90 min on a 15 rpm shaker at room temperature. It was rewashed three times carefully, followed by 90 min incubation of 100 µL of recombinant Zika NS1 antigen (BF-NS1-6309,

Biofront Technologies, Tallahassee, Florida, USA) with different concentrations spiked in PBS at room temperature. One hundred microliters of HRP-labeled anti-Zika NS1 antibody (BF-1125-36-HRP, Biofront Technologies, Tallahassee, Florida, USA) with 1:1000 dilution buffer (PBS containing 0.02% Tween 20, 3% Bovine Serum Albumins (BSA)) was added and incubated for 60 min on a 15 rpm shaker at room temperature followed by three washes with PBS. Blue color development was carried out using a 3,3′,5,5′-tetramethylbenzidine (TMB) substrate (PI34028, Fisher Scientific, Hampton, New Hampshire, USA) by incubating for 10 min in a dark place, and 2M H_2SO_4 was used to stop the color development. The absorbance was measured at 450 nm using a SpectraMax Gemini™ XPS/EM Microplate Reader (Molecular Devices, San Jose, CA, USA).

2.2. Microfluidic Chip Design

We fabricated a three-layer microchip using a cheaper non-lithographic method by laser cutting (Universal Laser) polymethylmethacrylate (PMMA) and double-sided adhesive (DSA) materials using an optimized design previously published [20]. The top layer (750 µm thick) has 0.4 mm diameter inlets and outlets to fit the pipette tips to allow easy sample loading. The middle layer (1.5 mm) contains reservoirs for all the mineral oil and aqueous solution, and the bottom layer (750 µm thick), which is solid, works as a base for the microchip. All the layers are assembled by using DSA films (Supplementary Figure S2a,b), and then pressed uniformly to remove all the bubbles by a bench vise (Home Depot).

2.3. Conjugation of Magnetic Beads with Capture Antibody

The capture antibody was first biotinylated using a type B fast biotin conjugation kit (ab201796) obtained from Abcam. One hundred microliters of Zika NS1 Mab (1225-36, Biofront Technologies) was modified by 10 µL of biotin-modified reagent and agitated very gently. The antibody mixer was added to the lyophilized biotin vial (100 µg), followed by 15 min incubation at room temperature. Ten microliters of Quencher reagent was added to stop the biotin reaction and was incubated for 4 min. Magnetic particles with an average diameter of 1 µm coated with neutravidin (GE Healthcare, Chicago, IL, USA) show higher affinity surface functionalization for biotinylated antibodies. Four hundred microliters of neutravidin-coated magnetic beads with 3500–4500 picomole/mg binding capacity were washed with 4 mL of PBS in a 5 mL Eppendorf Protein LoBind tube (14-282-304, Fisher Scientific). A magnetic stand was used for attracting all the magnetic beads creating a pellet on the magnetic stand side tube wall. The supernatant was discarded using a pipette and substituted by the same amount of PBS and mixed gently with a pipette to wash the magnetic beads. The process was repeated two times. For Zika NS1 capturing, 4 mL of 25 µg/mL biotinylated Zika NS1 Mab (1225-36, Biofront Technologies) was conjugated with the neutravidin-coated magnetic beads. To create a magnetic bead–biotin–capture antibody conjugation, the mixer was incubated overnight on a shaker (15 rpm) at 4 °C. After incubation; the magnetic bead conjugation was washed three times following the earlier mentioned steps to remove boundless Zika NS1 antibodies. One hour of blocking was done by using SuperBlock T20 (PBS) Blocking Buffer (37517, Fisher Scientific) to functionalize the magnetic beads at room temperature followed by washing twice using 4 mL PBS. After rinsing the magnetic beads, the conjugation was resuspended in 4 mL PBS and stored at 4 °C for future use.

2.4. M-ELISA on 96-Well Plate

Before using the magnetic bead–antibody solution, the beads were vortexed for 2–3 s to make a homogeneous solution. Thirty microliters of (1 mg/mL) magnetic beads conjugated with capture antibody was added on a conical-bottom 96-well plate (12-565-215, Fisher Scientific), and then all supernatant was isolated and discarded with the help of a 96-well-plate magnetic separator stand. One hundred microliters of Zika NS1 antigen spiked in plasma was added on to the wells and mixed gently to create a homogeneous solution followed by a 45 min incubation at room temperature on a shaker at 15 rpm speed. After that, the beads were aggregated by applying them onto the magnetic separator for 30–40 s to allow the magnetic beads to create a pellet on the sidewall of the wells. The liquid was discarded by a pipette and replaced with 200 µL of PBS to wash the beads. Beads were washed three times carefully to remove all the uncaptured antigen. One hundred microlites of HRP-labeled anti-Zika NS1 antibody (BF-1125-36-HRP, Biofront Technologies Tallahassee, Florida, USA) with 1:1000 dilution buffer (PBS containing 0.02% Tween 20, 3% BSA) was added and mixed slowly and incubated for 15 min on a 15 rpm shaker at room temperature. The beads were again washed three times by following the procedure mentioned above. One hundred microliters of TMB substrate was added, mixed, and incubated for 90 s in a dark place to generate a blue color. Color development was stopped by using 100 µL of 2M H_2SO_4. Beads were isolated with the help of a magnetic separator, and the liquid was transferred to a new well and absorbance was measured at 450 nm by using a SpectraMax Gemini™ XPS/EM Microplate Reader (Molecular Devices, San Jose, CA, USA).

2.5. Automated M-ELISA on Chip

An Arduino-controlled in-house-built magnetic actuation platform [20] was used to facilitate the automation of the developed assay (Figure 1a,b). The platform consisted of two parts including an Arduino controlling unit and the 3D printed platform, which could accommodate the microfluidic chip. The control unit housed the stepper motor driver (Pololu A4988) to regulate the bidirectional motor movement through a linear slide enclosed with two magnets (Supplementary Movie S1). The controlled unit was commanded by a gcode script set using a computer interface and powered by an external power supply.

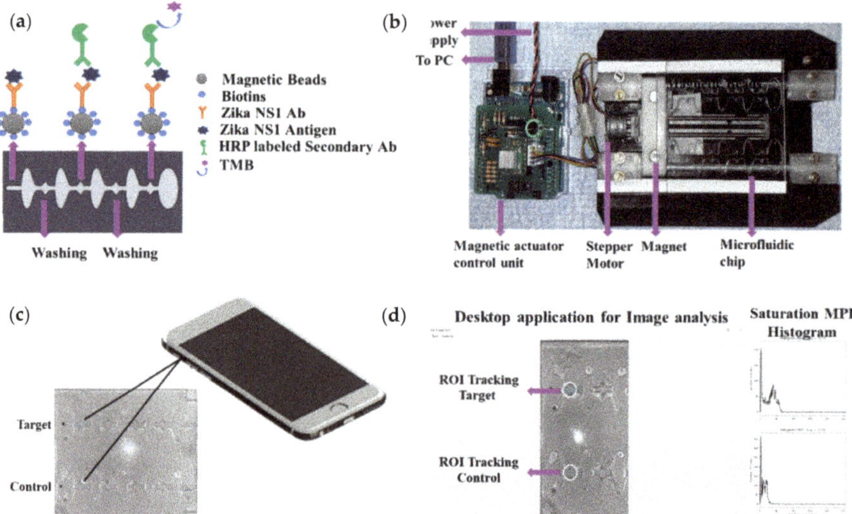

Figure 1. Graphical representation of the microfluidic sandwich enzyme-linked immunosorbent assay (ELISA) inside of a microfluidic chip. (**a**) Captured antigen is loaded inside the chip and moved through a washing buffer. The antigen-captured beads are labeled with the HRP-conjugated secondary antibody in the next chamber. After moving through a washing buffer chamber again, blue color was developed by reacting with a color-generation substrate. The reaction was stopped by moving the beads to a retention chamber. (**b**) A magnetic actuation platform containing an Arduino-controlled stepper motor unit. The 3D printed platform facilitates stepper motor housing allowing the user to move the magnets according to the command set up using a PC. The platform also holds the microfluidic chip just above the magnets. (**c**) Video frames of developed color on the chip are directly captured by a smartphone. (**d**) Region-of-interest (ROI) tracking using a desktop application and a histogram plot of the saturation maximum pixel intensity (MPI) of the developed color by the microfluidic magnetic enzyme linked immunosorbent assay (M-ELISA) on chip.

The prefabricated three-layer microfluidic chip was first loaded with all the reagents. First, 40 μL of (1 mg/mL) magnetic beads conjugated with capture antibody were taken into a 0.5 mL protein LoBind PCR tube, and then the supernatant was isolated with the help of a magnet and discarded by a pipette. One hundred microliters of Zika NS1 antigen diluted 2000-62.5 ng/mL concentration was added to the PCR tube and mixed gently. Then the PCR tube was incubated for 10 minutes at room temperature on a 15 rpm shaker to allow the beads to capture the antigen. In the meantime, the other reagents were loaded into the microfluidic chip with the help of a pipette. The loading steps in brief were as follows (Supplementary Figure S1), first both of the washing chambers (1,2) were filled with PBS, then HRP-labeled anti-Zika NS1 antibody (1:500 diluted in PBS containing 0.02% Tween 20, and 3% BSA) was loaded in chamber 3 on the chip followed by TMB substrate loading on the color-generation chamber, chamber 4. The TMB substrate was loaded in a dark place, and, after loading, the chamber was covered by an opaque tape to avoid oxidation with the reaction of normal light. After that, mineral oil (Sigma Aldrich, St. Louis, Missouri, USA) with a viscosity of 15 cst was loaded in chambers 5, 6, 7 and 8. After the Zika NS1 capture in the PCR tube, magnetic beads were isolated, and the supernatant was removed. Thirty microliters of PBS was replaced on the tube and mixed gently with the beads. Then the beads solution was loaded into chamber 9. Finally, chip loading was completed by loading mineral oil in chamber 10. The full loading process took 6–7 min. After loading the chip, it was placed on the magnetic actuation platform to perform washing, labeling with HRP-conjugated anti-Zika NS1 antibody, and color development, automatically (Figure 1a). The M-ELISA on-chip steps, in brief, were as follows, the magnets placed under the chip, which was enclosed by stepper motor, first moved inside chamber 9 for 15 s to accumulate all the beads and create a pellet. Then the beads were washed in chamber 1 for one minute to remove any uncaptured NS1 proteins. After washing, the beads were moved to chamber 3, where the beads were probed with HRP-labeled anti-Zika NS1 antibody. To remove all the nonspecifically bounded HRP-labeled anti-Zika NS1, the beads were washed again for one minute in chamber 2. The blue color was generated in the chamber 4 by reacting with TMB substrate for 30 s. The fully automated ELISA process took approximately 19–20 min including the sample preparation and loading.

2.6. Image Acquisition and Analysis

The microfluidic chip was placed on a white paper on a flat surface instantly after the completion of the M-ELISA. The covering tape was removed as quickly as possible, and a 30 s video was recorded using iPhone Xs cellphone (Figure 1c). The camera flashlight was on during the video recording (30 fps) to avoid external light interference and light variation from external sources. The white paper under the chip provided uniform background conditions. An OpenCV- and Python-based image processing semiautomated windows desktop application was used for image analysis (Figure 1d). First, the video was segregated frame by frame using a MATLAB (MathWorks, Natick, MA, USA) code and each 30th frame (equal to 1 second) was considered to select 30 images from the total 30 s video. Each image was analyzed by the desktop application to quantify the saturation maximum pixel intensity (MPI) in sequence. To analyze the intensity of color change, we manually selected the region of interest (ROI) using the desktop application. The desktop application facilitates the user to draw, drag, and drop a circle on top of the ROI of the analyzed images (Figure 1d). The arithmetic mean of the saturation channel was calculated during the colorimetric analysis. In this analysis, the ROI was chosen based on the portion having maximum saturation intensity for both sample and control channels. To set the sensitivity of the assay, the mean saturation pixel value of negative control (Dengue 2 NS1) as a result of M-ELISA on a 96-well plate was considered as the basis (mean ±3 standard deviation).

3. Results and Discussion

Conventional gold standard sandwich ELISA was carried out on a 96-well plate with multiple dilutions of Zika NS1 antigens (78.5 pg/mL–80 ng/mL) to validate the antibody and the target binding. Results clearly show that the anti-Zika NS1 monoclonal antibody was binding as low as 78 pg/mL of

Zika NS1 when it was spiked in PBS (Figure 2a) with a higher correlation ($R^2 \geq 0.9929$). In this study, we used Dengue 2 NS1 antigen as negative control and we also performed standard sandwich ELISA to further confirm that there was no cross reactivity/false-positive result shown by the anti-Zika NS1 monoclonal antibody (Supplementary Figure S3).

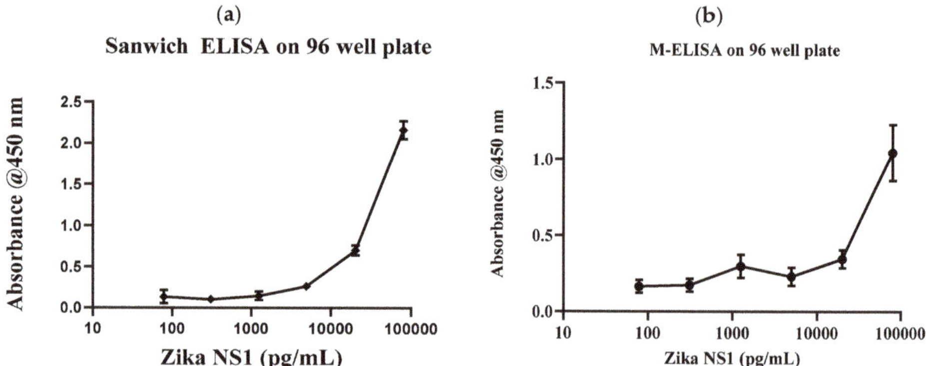

Figure 2. (**a**) Sandwich ELISA assay results using the recombinant Zika nonstructural protein 1 (NS1) antigen spiked on PBS buffer on 96-well plate as the Zika NS1 monoclonal antibody is used as a capture agent. Anti-Zika NS1 monoclonal antibody-HRP with a dilution factor of 1:1000 utilized to react with TMB calorigenic substrate to develop color. (**b**) M-ELISA assay on a 96-well plate showing spiking recombinant Zika NS1 on whole plasma. A 1:500 dilution factor of HRP-labeled anti-Zika NS1 was used and reacted with TMB to generate color. In both cases, color development was stopped by using H_2SO_4 and absorbance was measured at 450 nm using a SpectraMax Gemini™ XPS/EM Microplate Reader (Molecular Devices, USA). Error bars are ±SD.

After validating the Ag–Ab reaction, the anti-Zika NS1 monoclonal antibody was conjugated with superparamagnetic beads using a commercially available biotin conjugation kit. To confirm the magnetic bead–biotin–antibody binding, we performed a zeta potential analysis. A change in the negative surface charge indicates that the beads were successfully conjugated with antibodies. Supplementary Figure S4 shows that the surface charge of the magnetic beads was −5.63 mV and it was changed to −21.4 mV after the beads–antibody conjugation. With the antibody-conjugated beads, we developed and optimized a protocol for the M-ELISA to be performed on a 96-well plate. Again, different concentrations of Zika NS1 protein (78.5 pg/mL–80 ng/mL) spiked in whole plasma were used for the standard curve readouts; whereas Dengue 2 NS1 antigen spiked in plasma was used as a negative control. We also optimized the dilution factor of HRP-labeled anti-Zika NS1 antibody for the M-ELISA (not reported) to reduce the background readout. The absorbance reading was taken by the SpectraMax Gemini™ XPS/EM Microplate Reader. A picture of the 96-well plate was taken with camera flashlight on. After extracting the saturation MPI of the taken image, it confirmed that both the M-ELISA standard curve (Figure 2b), as well as the desktop application, generated the saturation MPI standard curve (Figure 3a) and presented comparable results. The observed detection limit of Zika NS1 in M-ELISA on 96-well plate was 78 pg/mL ($R^2 \geq 0.9929$). This result indicates that the M-ELISA is also consistent like conventional sandwich ELISA. Therefore, the measurement of the saturation MPI to calculate the concentration of Zika NS1 is also reliable.

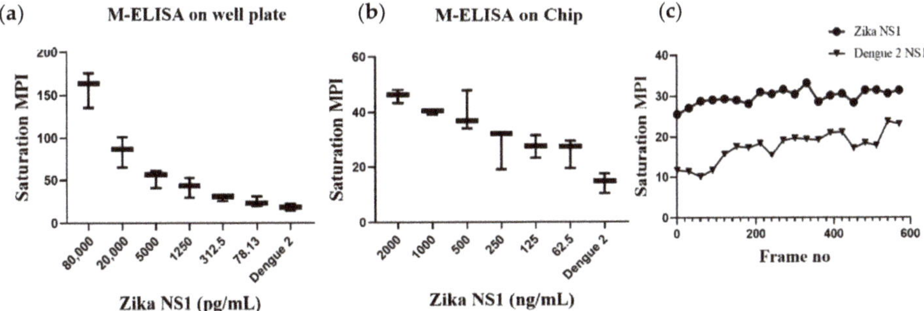

Figure 3. Standard curve of saturation MPI (**a**) for a captured cell phone image of the 96-well plate after performing M-ELISA on the 96-well plate and (**b**) M-ELISA on chip by computing the developed color intensity of the segmented frames of a captured video. An OpenCV- and Python-based semiautomated desktop application were used to select the ROI to calculate the saturation MPI. The use of the cell phone quantitation method on an M-ELISA showing a limit of detection of 78 pg/mL ($R^2 \geq 0.9561$) and 62.5 ng/mL ($R^2 \geq 0.9929$) on 96-well plate and microfluidic chip, respectively. Error bars are ±SD. (**c**) Comparison of saturation MPI change over time, quantified by analyzing a 20 s video (30 fps) just after completing the M-ELISA on chip. The full video was segmented frame by frame, and each 30th frame was analyzed to compare the saturation MPI change over frames (time).

To further reduce the required reagent and assay timing for performing the M-ELISA we used a previously reported [20] Arduino-controlled magnetic actuation platform (Figure 1b).

To fabricate the three layers of microfluidic chip to accommodate different reagents, we chose 0.75 mm thick PMMA for the top and bottom layers and a 1.5 mm middle layer (Supplementary Figure S2). The total volume of the chip was 907 µL, and it contained a total of 10 chambers. Our designed chip has been optimized such that (Supplementary Figure S2c) it will have the highest bead-to-magnet attraction, low bead loss while moving from one chamber to another chamber, and reliable quantification based on previously published work [20,21]. Three differently shaped chambers were used while designing the microfluidic chips. The cylindrical chamber is considered as the bead aggregation chamber and the magnetic actuation starts from this chamber. In this chamber, the magnet oscillates two times for approximately 16 seconds to facilitate creation of the bead pellet. The four diamond-shaped chambers contain washing buffer, HRP-labeled anti NS1 antibody, and TMB substrates. All the elliptical chambers contained mineral oil which basically acts as a physical barrier between each aqueous reagent. The shapes are designed and optimized in a way that provides minimum surface tension and minimum meniscus effect while the magnetic bead moves from one phase to another linearly.

After validating target Ag–Ab binding by conventional sandwich ELISA and M-ELISA (Figure 2a,b), the application of the M-ELISA on the chip was performed using the automated platform. The assay run time was optimized to 9 min and 20 min (Table 1) including sample preparation and loading. Until now the Zika NS1 levels in Zika-infected patients were still largely unknown/variable, whereas approximately 15 µg/mL Dengue NS1 can be found for Dengue-infected patient's serum after two days of infection [12,22]. Based on that, we have analyzed 2000.0, 1000.0, 500.0, 250.0, 125.0, and 62.5 ng/mL Zika NS1 spiking on plasma to perform M-ELISA on-chip (Figure 3b). The observed detection limit for the M-ELISA on-chip was 62.5 ng/mL.

Table 1. Time comparison of sandwich ELISA, M-ELISA, and M-ELISA on chip.

	Steps	Sandwich ELISA	M-ELISA	M-ELISA on Chip
Time consumption	Coating	Overnight	Pre-prepared	Pre-prepared
	Antigen capture	1.5 h	45 min	10 min (outside chip)
	Blocking	1.5 h	–	–
	Secondary antibody labeling	1 h	15 min	5 min
	Development	10 min	1.5 min	30 s
	Washing	20 min	10 min	2 min
	Beads collection and moving	–	–	1 min
	Total Time	~4.5 h	~72 min	~9 min
Sample consumption (μL)	Coating	100	Pre-prepared	Pre-prepared
	Antigen Capture	100	100	~30
	Blocking	200	Pre-prepared	Pre-prepared
	Secondary antibody labeling	100	100	~60
	Development	100	100	~60
	H_2SO_4	100	100	–
	Washing	1500	900	~120
	Separation (oil)	–	–	~464
	Retention (oil)	–	–	~173
	Total reagent	2200	1300	~907
	Limit of detection	78 pg/mL	78 pg/mL	62.5 ng/mL

To the best of our evidence, this is the first reported automated sandwich ELISA technique for detecting Zika NS1 antigen. Our goal was to develop and confirm the sandwich ELISA with superparamagnetic beads on a microfluidic chip. The non-lithographic process of chip fabrication has been reported previously [21,23–25] for inexpensive, disposable, and robust applications for various analyte detections in resource-limited settings. Supplementary Table S1 shows the manufacturing cost for each developed microfluidic chip and the cost per assay. We used smartphone captured video and an in-house-developed desktop application to calculate the color development by TMB. Video capture instead of a single-picture capture provides the flexibility of choosing a large set of images taken over a period of time [26]. In this study, video capture plays an important role as there was no stopping solution after the TMB reaction. After the completion of beads' movement and color development, the color was changing until saturation. Since saturation can measure the intensity of the color, it is a suitable measure for concentration-dependent colorimetric assays, and it has been reported previously [26,27]. Saturation represents the amount of intensity of color in an image and can be represented from 0 to 255 in binary scale. The intensity of color development is correlated with the concentration-dependent colorimetric assays. While considering the image for calculating the saturation MPI for both samples and control, each 30th frame was considered to record the color development. Unlike conventional sandwich ELISA, our M-ELISA chip method contains no stopping solution. As a result of this, the developed color intensity increases over the video capturing time. Figure 3c shows the relation between both samples and control color change over time. While quantifying the saturation MPI, we consider the mean ±3 standard deviation of the control's saturation MPI of M-ELISA on a 96-well plate as a basis to determine the sensitivity irrespective of any specific frame or time of the captured video.

Recently, very few studies have been published for developing detection assays targeting ZIKV NS1 antigen. A double antibody sandwich ELISA-based colorimetric assay was able to detect it as low as 120 ng/mL [6]. Another sensing platform using a graphene biosensor was able to detect 500 ng/mL spiked in a 10-fold diluted serum [28]. An antigen–antibody-based, rapid paper-based diagnostic assay has been reported to have a sensitivity of about 20 ng/mL for ZIKV NS1 [29]. Following an alternative approach, our developed assay was able to detect as low as 62.5 ng/mL in whole plasma, which is comparable to other currently existing or developed techniques.

Moreover, the developed assay is suitable for detecting the Zika NS1 antigen without the help of costly specialized instruments (e.g., a microplate reader), which in turn reduces cost per assay to below $2 (USD) (Supplementary Table S1). The unique features for the developed assays are automation, rapid assay turnout, and result readout by cell-phone-based video capture. Sample handling and loading do not require any skilled personnel. The developed device design is highly scalable and microfluidic chips can be designed to run up to 12 assays in one go by further optimization. To further reduce assay loading time, preloaded devices can be designed and tested. Considering the throughput against 96-/384-well plate-based conventional sandwich ELISA, our developed platform has lower throughput but considering other factors especially portability, cost, and applicability it is suitable for developing countries where resources are limited, and a smaller number of tests are required per day. The developed assay fulfills ASSURED criteria [30] (i.e., affordable, sensitive, specific, user-friendly, rapid, equipment-free, and deliverable) for being applicable to the POC.

4. Conclusions

We developed an automated microfluidic assay integrated with smartphone that can obtain results in ~10 minutes, which can be beneficial in resource-limited settings. There were some shortcomings and further areas for our device to expand on. The sensitivity of our device is much higher than that of other point-of-care devices with a detection limit of 62.5 ng/mL. Further improvement in this area can result in great potential for this device to be clinically used in areas around the world. However, the cost effectiveness, automation, and time effectiveness compared to traditional ELISA's can be beneficial in regions that struggle with poverty and lack of resources. The utilization of smartphone video and the in-house software can be operated by low-skilled workers. This, in turn, can further the accessibility of the developed device and allow anyone to record and interpret results. Without the use of high-cost specialized instruments, the developed chip can be kept at a low cost. These results show potential that the platform would be useful for the POC diagnosis/screening of Zika virus infection patients and can be used in resource-limited settings.

Supplementary Materials: The following are available online at http://www.mdpi.com/2075-4418/10/1/42/s1, Table S1: Material and reagents cost of the disposable microfluidic chip, Figure S1: Reagent Loading sequence inside the microfluidic chip, Figure S2: Chip design, fabrication, and assembly, Figure S3: Sandwich ELISA assay for Dengue 2 NS1 with the capture antibody for Zika NS1 for detection in binding buffer, Figure S4: Zeta potential measurement of antibody coated beads, Video S1: Sample reagent loading and test run.

Author Contributions: Conceptualization, M.A.K. and W.A.; methodology, M.A.K., H.Z., M.S, and W.A.; software, M.A.K.; validation, M.A.K., H.Z. and M.S.; formal analysis, M.A.K., H.Z. and, W.A.; investigation, W.A.; resources, M.A.K. and W.A.; data curation, M.A.K. and W.A.; writing—original draft preparation, M.A.K. and H.Z.; writing—review and editing, M.A.K., H.Z., M.S., and W.A.; visualization, M.A.K.; supervision, W.A.; project administration, W.A.; funding acquisition, W.A. All authors have read and agreed to the published version of the manuscript.

Funding: This research was funded by National Institute of Health (NIH) grant number R15AI127214, and R56AI138659, and Florida Department of Health Zika grant 7ZK10. The APC was funded by National Institute of Health (NIH) grant number R56AI138659.

Acknowledgments: We acknowledge research support from NIH R15AI127214, NIH R56AI138659, Florida Department of Health ZIKA grant (FDOH) 7ZK10, and support from the College of Engineering and Computer Science, Florida Atlantic University, Boca Raton, FL.

Conflicts of Interest: The authors declare no conflict of interests.

References

1. Sikka, V.; Chattu, V.K.; Popli, R.K.; Galwankar, S.C.; Kelkar, D.; Sawicki, S.G.; Stawicki, S.P.; Papadimos, T.J. The Emergence of Zika Virus as a Global Health Security Threat: A Review and a Consensus Statement of the INDUSEM Joint Working Group (JWG). *J. Glob. Infect. Dis.* **2016**, *8*, 3–15. (In English) [PubMed]
2. Herrada, C.A.; Kabir, M.; Altamirano, R.; Asghar, W. Advances in Diagnostic Methods for Zika Virus Infection. *J. Med. Devices* **2018**, *12*, 4. [CrossRef] [PubMed]
3. PAHO; WHO. Cases of Zika Virus Disease. Available online: http://www.paho.org/data/index.php/en/?option=com_content&view=article&id=524&Itemid= (accessed on 3 October 2019).
4. WHO. Zika Epidemiology Update. Available online: https://www.who.int/emergencies/diseases/zika/zika-epidemiology-update-july-2019.pdf?ua=1 (accessed on 8 October 2019).
5. Musso, D.; Roche, C.; Robin, E.; Nhan, T.; Teissier, A.; Cao-Lormeau, V.-M. Potential sexual transmission of Zika virus. *Emerg. Infect. Dis.* **2015**, *21*, 359. [CrossRef] [PubMed]
6. Zhang, L.; Du, X.; Chen, C.; Chen, Z.; Zhang, L.; Han, Q.; Xia, X.; Song, Y.; Zhang, J. Development and characterization of double-antibody sandwich ELISA for detection of zika virus infection. *Viruses* **2018**, *10*, 634. [CrossRef]
7. World Health Organization. *Laboratory Testing for Zika Virus Infection: Interim Guidance*; World Health Organization: Geneva, Switzerland, 2016.
8. Landry, M.L.; George, K.S. Laboratory diagnosis of Zika virus infection. *Arch. Pathol. Lab. Med* **2016**, *141*, 60–67. [CrossRef]
9. Matheus, S.; Boukhari, R.; Labeau, B.; Ernault, V.; Bremand, L.; Kazanji, M.; Rousset, D. Specificity of dengue NS1 antigen in differential diagnosis of dengue and Zika virus infection. *Emerg. Infect. Dis.* **2016**, *22*, 1691. [CrossRef]
10. Lee, K.H.; Zeng, H. Aptamer-based ELISA assay for highly specific and sensitive detection of Zika NS1 protein. *Anal. Chem.* **2017**, *89*, 12743–12748. [CrossRef]
11. Rong, Z.; Wang, Q.; Sun, N.; Jia, X.; Wang, K.; Xiao, R.; Wang, S. Smartphone-based fluorescent lateral flow immunoassay platform for highly sensitive point-of-care detection of Zika virus nonstructural protein 1. *Anal. Chim. Acta* **2019**, *1055*, 140–147. [CrossRef]
12. Sánchez-Purrà, M.; Carré-Camps, M.; de Puig, H.; Bosch, I.; Gehrke, L.; Hamad-Schifferli, K. Surface-enhanced raman spectroscopy-based sandwich immunoassays for multiplexed detection of zika and dengue viral biomarkers. *ACS Infect. Dis.* **2017**, *3*, 767–776. [CrossRef]
13. Wang, S.M.; Sekaran, S.D. Evaluation of a commercial SD dengue virus NS1 antigen capture enzyme-linked immunosorbent assay kit for early diagnosis of dengue virus infection. *J. Clin. Microbiol.* **2010**, *48*, 2793–2797. [CrossRef]
14. Cecchetto, J.; Fernandes, F.C.; Lopes, R.; Bueno, P.R. The capacitive sensing of NS1 Flavivirus biomarker. *Biosens. Bioelectron.* **2017**, *87*, 949–956. [CrossRef] [PubMed]
15. Chen, Y.; Wang, Y.; Liu, L.; Wu, X.; Xu, L.; Kuang, H.; Li, A.; Xu, C. A gold immunochromatographic assay for the rapid and simultaneous detection of fifteen β-lactams. *Nanoscale* **2015**, *7*, 16381–16388. [CrossRef] [PubMed]
16. Song, S.; Liu, N.; Zhao, Z.; Njumbe Ediage, E.; Wu, S.; Sun, C.; De Saeger, S.; Wu, A. Multiplex lateral flow immunoassay for mycotoxin determination. *Anal. Chem.* **2014**, *86*, 4995–5001. [CrossRef] [PubMed]
17. Zhang, L.; Huang, Y.; Wang, J.; Rong, Y.; Lai, W.; Zhang, J.; Chen, T. Hierarchical Flowerlike Gold Nanoparticles Labeled Immunochromatography Test Strip for Highly Sensitive Detection of Escherichia coli O157:H7. *Langmuir* **2015**, *31*, 5537–5544. [CrossRef] [PubMed]
18. Choi, J.R.; Yong, K.W.; Tang, R.; Gong, Y.; Wen, T.; Yang, H.; Li, A.; Chia, Y.C.; Pingguan-Murphy, B.; Xu, F. Lateral Flow Assay Based on Paper–Hydrogel Hybrid Material for Sensitive Point-of-Care Detection of Dengue Virus. *Adv. Healthc. Mater.* **2017**, *6*, 1600920.
19. Theillet, G.; Rubens, A.; Foucault, F.; Dalbon, P.; Rozand, C.; Leparc-Goffart, I.; Bedin, F. Laser-cut paper-based device for the detection of dengue non-structural NS1 protein and specific IgM in human samples. *Arch. Virol.* **2018**, *163*, 1757–1767. [CrossRef]
20. Coarsey, C.; Coleman, B.; Kabir, M.A.; Sher, M.; Asghar, W. Development of a flow-free magnetic actuation platform for an automated microfluidic ELISA. *RSC Adv.* **2019**, *9*, 8159–8168. [CrossRef]

21. Wang, S.; Tasoglu, S.; Chen, P.Z.; Chen, M.; Akbas, R.; Wach, S.; Ozdemir, C.I.; Gurkan, U.A.; Giguel, F.F.; Kuritzkes, D.R.; et al. Micro-a-fluidics ELISA for Rapid CD4 Cell Count at the Point-of-Care. *Sci. Rep. Artic.* **2014**, *4*, 3796. [CrossRef]
22. Steinhagen, K.; Probst, C.; Radzimski, C.; Schmidt-Chanasit, J.; Emmerich, P.; van Esbroeck, M.; Schinkel, J.; Grobusch, M.P.; Goorhuis, A.; Warnecke, J.M. Serodiagnosis of Zika virus (ZIKV) infections by a novel NS1-based ELISA devoid of cross-reactivity with dengue virus antibodies: A multicohort study of assay performance, 2015 to 2016. *Eurosurveillance* **2016**, *21*, 50. [CrossRef]
23. Asghar, W.; Sher, M.; Khan, N.S.; Vyas, J.M.; Demirci, U. Microfluidic Chip for Detection of Fungal Infections. *ACS Omega* **2019**, *4*, 7474–7481. [CrossRef]
24. Sher, M.; Asghar, W. Development of a multiplex fully automated assay for rapid quantification of CD4+ T cells from whole blood. *Biosens. Bioelectron.* **2019**, *142*, 111490. [CrossRef] [PubMed]
25. Wang, S.; Zhao, X.; Khimji, I.; Akbas, R.; Qiu, W.; Edwards, D.; Cramer, D.W.; Ye, B.; Demirci, U. Integration of cell phone imaging with microchip ELISA to detect ovarian cancer HE4 biomarker in urine at the point-of-care. *Lab-on-a-Chip* **2011**, *11*, 3411–3418. [CrossRef] [PubMed]
26. Coleman, B.; Coarsey, C.; Kabir, M.A.; Asghar, W. Point-of-care colorimetric analysis through smartphone video. *Sens. Actuators B Chem.* **2019**, *282*, 225–231. [CrossRef] [PubMed]
27. Coleman, B.; Coarsey, C.; Asghar, W. Cell phone based colorimetric analysis for point-of-care settings. *Analyst* **2019**, *144*, 1935–1947. [CrossRef]
28. Afsahi, S.; Lerner, M.B.; Goldstein, J.M.; Lee, J.; Tang, X.; Bagarozzi, D.A., Jr.; Pan, D.; Locascio, L.; Walker, A.; Barron, F. Novel graphene-based biosensor for early detection of Zika virus infection. *Biosens. Bioelectron.* **2018**, *100*, 85–88. [CrossRef]
29. Bosch, I.; De Puig, H.; Hiley, M.; Carré-Camps, M.; Perdomo-Celis, F.; Narváez, C.F.; Salgado, D.M.; Senthoor, D.; O'Grady, M.; Phillips, E. Rapid antigen tests for dengue virus serotypes and Zika virus in patient serum. *Sci. Transl. Med.* **2017**, *9*, eaan1589. [CrossRef]
30. Peeling, R.W.; Mabey, D.; Herring, A.; Hook, E.W. Why do we need quality-assured diagnostic tests for sexually transmitted infections? *Nat. Rev. Microbiol.* **2006**, *4*, 909. [CrossRef]

diagnostics 

Article

Comparison of Night, Day and 24 h Motor Activity Data for the Classification of Depressive Episodes

Julieta G. Rodríguez-Ruiz [1,†], Carlos E. Galván-Tejada [1,*,†], Laura A. Zanella-Calzada [2,†], José M. Celaya-Padilla [3], Jorge I. Galván-Tejada [1], Hamurabi Gamboa-Rosales [1], Huizilopoztli Luna-García [1] and Rafael Magallanes-Quintanar [1] and Manuel A. Soto-Murillo [1]

[1] Unidad Académica de Ingeniería Eléctrica, Universidad Autónoma de Zacatecas, Jardín Juarez 147, Centro, Zacatecas 98000, Mexico; jr.ruiz68@uaz.edu.mx (J.G.R.-R.); gatejo@uaz.edu.mx (J.I.G.-T.); hamurabigr@uaz.edu.mx (H.G.-R.); hlugar@uaz.edu.mx (H.L.-G.); tiquis@uaz.edu.mx (R.M.-Q.); alejandro.somu@uaz.edu.mx (M.A.S.-M.)

[2] LORIA, Université de Lorraine, Campus Scientifique BP 239, 54506 Nancy, France; laura.zanella-calzada@univ-lorraine.fr

[3] CONACYT, Universidad Autónoma de Zacatecas, Jardín Juarez 147, Centro, Zacatecas 98000, Mexico; jose.celaya@uaz.edu.mx

* Correspondence: ericgalvan@uaz.edu.mx; Tel.: +52-492-5440968

† These authors contributed equally to this work.

Received: 30 December 2019; Accepted: 17 February 2020; Published: 17 March 2020

Abstract: Major Depression Disease has been increasing in the last few years, affecting around 7 percent of the world population, but nowadays techniques to diagnose it are outdated and inefficient. Motor activity data in the last decade is presented as a better way to diagnose, treat and monitor patients suffering from this illness, this is achieved through the use of machine learning algorithms. Disturbances in the circadian rhythm of mental illness patients increase the effectiveness of the data mining process. In this paper, a comparison of motor activity data from the night, day and full day is carried out through a data mining process using the Random Forest classifier to identified depressive and non-depressive episodes. Data from Depressjon dataset is split into three different subsets and 24 features in time and frequency domain are extracted to select the best model to be used in the classification of depression episodes. The results showed that the best dataset and model to realize the classification of depressive episodes is the night motor activity data with 99.37% of sensitivity and 99.91% of specificity.

Keywords: motor activity; depression; depressive episodes; data mining; random forest; night

---

1. Introduction

Wearable systems have been extensively used in healthcare field for several years. Physical medicine and rehabilitation were the first disciplines to venture into the implementation of these devices, in order to monitor the physical activity of the individual in pursuance of better disease diagnosis or patient ailment rehabilitation [1]. With the introduction of Internet of Things (IoT), wearable devices have been used to collect data from patients, not only for the detection of motor activity, but also for measuring blood pressure, heartbeat and even glucose level. This has allowed patients to be monitored any time and anywhere [2]. However, although the sensors are placed in the human body in the most normal and natural way possible to sense its activity, public data acquired from this type of sensing are still rare and not public [3].

Wearables have been used for the detection of depression according to their motor activity, since this represents a high indicator of the presence of this disease. Retardation or decrease in activities is the main feature for patients suffering from depression, for that reason, collect data from this condition could present good results that could be useful in different applications [4].

According to the World Health Organization (WHO) depression is the leading cause of disability [5], seven percent of people around the world suffer from Major Depressive Disorder (MDD), which causes the deterioration of the quality of life, the increase in medical costs and the death rate.

MDD is characterized by several syntopms as; loss of interest and pleasure in daily activities, sleep disorders, weight loss, suicide ideation, suicide attempts, among others. These symptoms must be present every day for at least two weeks to be diagnosed as depressive patients [6]. Depression is a treatable disease with a high level of efficacy using antidepressant medications and psychotherapy treatment, nevertheless for many patients being diagnosed can take months or even years to heal [7–9].

To diagnose or quantify the severity of the MDD, specialists use scales and manuals such as; the Hamilton Rating Scale for Depression written in 1960 [10], the Montgomery and Asberg Depression Rating Scale (MADRS) written in 1979 [11] or the Diagnostic and Statistical Manual of Mental Disorders (DSM). However, the use and interpretation of these methods depends largely on the ability of the specialist to determine the diagnosis [6,12–14]. A fact that discredits this type of methods is their lack of actualization and adaptation to the new advances and discoveries about the disease. In addition, these methods require the intervention of the patient, and in some cases the patients lie for any reason, causing the results not to be true and useful.

On the other hand, the use of data from monitoring depressive patients brings several benefits to medical services, mainly reduces the diagnosis and treatment time, improves the quality of life of patients and reduces medical costs [8].

Sensing motor activity arises as a favorable way for psychiatry and mental health to detect abnormal behaviors. Has been demonstrated that patients suffering from depression tend to reduce their daytime activity, and due to sleep disorders increase their nighttime activity [3]. In contrast, patients with bipolar disorder lead to an increase in their energy, however both scenarios presents a motor activity discrepancy from a healthy person. Therefore, circadian rhythm desynchronization is present in mental illness but is not well used for diagnosis or treatment monitoring yet [15].

The task of collect motor activity data can be accomplished using sensors like accelerometers, a combination of accelerometers, Global Positioning System (GPS), gyroscopes, inclinometers, magnetometers, etc. [1]. Nowadays, most of these technologies have very small dimensions, are cheep and easy to add in some specific devices or clothes, which facilitates usability and adaptability to everyday life. One way to replace these sensors could be using mobile phones, these devices have a big role in ubiquitous treatment, where the main idea is to avoid disturbances that the sensors or devices could generate on the patients and collect reliable data.

Once the data is collected the next step is to process it to recognize patterns and obtain some statistics or classifications. Sohrab Saeb et al [8] preset the relation of the regular clinical diagnosed and the sensor-based data from depression patients, they obtained an important result, in which a correlation between the GPS data and The Patient Health Questionnaire for depression (PHQ-9) scores was presented, this proves the relation between activity and depression [8].

In another work presented by Enrique Garcia-Ceja et al [3] a collected data from unipolar, bipolar and healthy control people was used to compare different machine learning algorithms and classify depressive and non-depressive signals, proving that through the use of machine learning techniques it is possible to classify between depressive and non-depressive people.

Machine learning is a set of algorithms that learn from the analyzed data to develop training models to classify that type of data. It allows among other applications, to make diagnoses or even predict some diseases [16]. These methods are commonly used in a data mining process that involves a series of steps related to each other and with the final objective of acquire valuable information.

Machine learning methods are increasingly used, EEG-based machine learning provides a non-invasive method to automatic diagnose MDD using algorithms like Linear Regression and Naive Bayes [17].

In this paper, a data mining process is carried out to classified depressive episodes using data collected during night time, day and full 24 h. The comparison between the classification using different data collected trough the time gives a better image of the disease and behavior of the patients with the diagnosis.

The structure of the paper is the Material and Methods, Results, Discussion and Conclusion sections, the Material and Methods section describes step by step the data mining process used to implement the classification of depressive and non-depressive episodes.

2. Materials and Methods

The data mining process [18] shown in Figure 1 is followed to classify depressive episodes. The first stage consists on the data collection, where the Depresjon dataset containing the information of depression episodes from patients is acquired. These data are submitted to a pre-processing step in order to clean, normalize and segment them in one hour lapses. Then, a feature extraction is applied, where a set of 24 features in the time and frequency domain are obtained for different stages of the day (day, night and full day). A feature selection based on a forward selection (FS) approach is subsequently performed to reduce the number of features and to avoid redundant or non-significant information. From the selected features, a classification step based on the random forest (RF) algorithm is applied to develop a series of generalized models to identify between healthy and depressed patients according to the motor activity. Finally, these models are validated by a statistical analysis.

Data Collection	Pre-processing	Features Extraction	Feature Selection	Classification	Validation
Description of the dataset used for the data mining process.	Cleaning and formatting the data to an specific structure.	Extraction of meaningful features from the data.	Selection of the best features that describes the data.	Implementation of machine learning algorithms to make a classification.	Assess how well models perform in the previous classification.

Figure 1. Data mining process used in this paper to classified depressive and non-depressive episodes.

2.1. Dataset Description

In this work, the Depresjon dataset is used to classify depressive episodes. It is comprised by the motor activity of 23 patients diagnosed as bipolar, unipolar depressive and bipolar I (all these labeled as condition), and 32 non-depressive control subjects.

The motor activity corresponds to a weighting voltage collected with an actigraph watch (Actiwatch, Cambridge Neurotechnology Ltd., England, model AW4) located in the right wrist, which records movements over 0.5g in a sampling frequency of 32 Hz. Some advantages of actigraphs accelerometers is that they are inexpensive and well-known activity trackers [19] and, in addition, they are easy to wear and allow collecting data from day and night [20].

The data structure is formed by different files. One set of files contains a csv file per each condition and control with their recorded motor activity, organized in three columns: timestamp (one minute intervals), date (date of measurement), activity (activity measurement from the actigraph watch). Also, a scores file is included, which provides information about every subject. This file includes the columns: number (contributor id), days (number of days of data collection), gender (1 or 2 for female or male), age, afftype (1: bipolar II, 2: unipolar depressive, 3: bipolar I), melanch (1: melancholia, 2: no melancholia), inpatient (1: inpatient, 2: outpatient), edu (education), marriage (1: married or cohabiting, 2: single), work (1: working or studying, 2: unemployed/sick leave/pension), madrs1 (MADRS score when measurement started), madrs2 (MADRS when measurement stopped).

Features describing the date, timestamp and if it is or not a weekend, are not taking into account for the classification.

The package with the dataset files and full description of data can be downloaded from http://doi.org/10.5281/zenodo.1219550 [15].

2.2. Pre-Processing

For the pre-processing stage, the next step are proposed. Since the total amount of data recorded for each subject is different, a new subset of data is extracted, adjusting the number of observations to be equal for each subject. Theh, from the new set of data, a segmentation is applied to form one hour data intervals. This segmentation allowed the classification of depressive episodes per hour.

Therefore, based on the hourly segmentation, three different subsets are constructed; night motor activity (from 21 to 7 h taking into account the sunrise standard hours) [21], day motor activity (from 8 to 20 h) and finally all day motor activity with the total day hours. The number of observations contained in each dataset is shown in Table 1. After separated the data into day, night and 24 h data were cleaned from missing data.

Table 1. Datasets created from Depresjon dataset.

Dataset	Observations
Day	14168
Night	11945
Full Day	26113

Finally, the last step for the pre-processing is the cleaning of the data by the elimination of missing data, represented as NA, and the standardization of the motor activity. This standardization center the data into the mean mark, allowing to know how far the signal is from the mean point. This standardization is calculated by Equation (1).

$$z_i = \frac{x_i - \bar{x}}{s},\tag{1}$$

where x_i is the actual point of the activity data, $\bar{x}$ is the mean of the total motor activity data and s is the standard deviation of the total motor activity data.

2.3. Feature Extraction

For each dataset, the 24 features shown in Table 2 are extracted. This process is based on similar works that extract features from an accelerometer signal [22–24].

From the total features extracted, ten are based on the time-domain, as shown in Table 2, referred to the data collected by the actigraph every minute.

To transform the time-domain data into frequency-domain, the fast Fourier transform (FFT) is applied, which can be calculated with Equation (2),

$$x(k) = \sum_{n=0}^{N-1} x(n) * e^{-j2\pi(x\frac{n}{N})},\tag{2}$$

where $x(n)$ represents each motor activity collected per minute on an hour, N represents the total observations on an hour lapse, k represents the current frequency taking values from 0 to $N-1$, and $x(k)$ represents the spectral components of the samples.

For this FFT process, the representation of the original signal in the frequency domain is computed using the discrete Fourier transformation (DFT). This representation is formed by complex numbers, eliminating the imaginary part of each number in the frequency-domain signal. For this transformation, it is needed to calculate the power spectral density (PSD), as shown in Equation (3),

$$P = \lim_{T\to\infty} \frac{1}{T} \int_0^T |x(k)|^2 \, dt,\tag{3}$$

where P represents the energy from the signal, T represents the length of the signal lapse and $x(k)$ represents the frequency-domain signal. The spectrum is normalized by the length of the signal.

The 14 remaining features are extracted from the PSD of the signal to obtain the best characterization of the signal.

Table 2. Features extracted for the day, night and full day datasets, in time and frequency domain.

Feature	Equation	Time Domain	Frequency Domain
Mean (μ)	$1/n \sum_{i=1}^{n} x_i$ [1]	•	•
Median	$x_{(n+1)/2}$	•	•
Standard deviation (SD, σ)	$\sqrt{(\sum_{i=1}^{n}(x_1 - \bar{x})^2/(n-1))}$	•	•
Variance	$1/n \sum_{i=1}^{n}(x_i - \mu)^2$	•	•
Kurtosis	μ_4/σ_4 [2]	•	•
Coefficient of Variance	σ/μ	•	•
Interquartile range	$Q_3 - Q_1$ [3]	•	•
Minimum	(*Maximum value*)	•	•
Maximum	(*Minimum value*)	•	•
Trimmed Mean	(*truncated mean*)	•	•
Spectral Density	(defined above)		•
Entropy	$-\sum_i p(x_i) log_2 p(x_i)$ [4]		•
Skewness	$\overline{\mu_3}$ [5]		•
Spectral Flatness	$(exp(1/N \sum_{n=0}^{N-1} lnx(n))/(1/N \sum_{n=0}^{N-1} x(n)$ [6]		•

[1]—n = total number of samples, x_i = actual sample. [2]—μ_4 = μ of the fourth moment, σ_4 = SD of the fourth moment. [3]—Q_3 = third quartile three, Q_1 = first quartile. [4]—$p_i(x_i)$ = probability of x_i. It represents the media uncertainty of a random variable. [5]—$\overline{\mu_3}$ = third standardized moment. [6]—$x(n)$ = magnitude of bin number n.

2.4. Feature Selection

The next step consists on reducing the dimension of the feature sets and selecting the best model for the description of the data. To accomplish the task, a FS approach is applied to the three sets of features (day, night and full day), using 70% of the data for the training of the model and the 30% remaining data for the testing of the model [25].

FS is implemented using the logistic regression (LR) classifier, since the nature of the data is binary (depressive, "1", and not depressive, "0", episode). Therefore, LR is used to model the selected features by FS for the classification of depressive episodes. For simplicity, each feature is labeled with a number, as shown in Table 3.

Table 3. Features extracted from the hourly motor activity segments.

Number	Time-Domain Feature	Number	Frequency-Domain Feature
0	Kurtosis	10	Kurtosis
1	Mean	11	Mean
2	Median	12	Median
3	SD	13	SD
4	Variance	14	Variance
5	Coefficient of variance	15	Coefficient of variance
6	Interquartil rank	16	Spectral density
7	Minimum	17	Interquartile rank
8	Maximum	18	Trim mean
9	Trim mean	19	Minimum
		20	Maximum
		21	Entropy
		22	Skewness
		23	Spectral flatness

2.5. Classification

For the classification stage, the RF algorithm is used to classify depressive and non-depressive episodes based on the features selected for each dataset, specifically using the best ranked features according to the previous step.

According to Phan Thanh Noi et al [26], the RF algorithm use has been increasing in the past few years because of its effectiveness. RF algorithm came to light in 2001 created by Leo Breiman et al [27], it is conformed by a combination of trees generated randomly and with different predictors each of them. This algorithm is a supervised technique where multiple decision trees are used to develop a forest. This forest is more robust if it is developed with more number of trees for the classification.

To classify an observation, the trees are generated in order to response questions with yes/no response, every tree bases the response in the features of the observation and responded to make a classification of the observation [28].

Generally, a leaf is used for the expansion of the construction of the tree in each step. At the end, from the decision trees built, they are merged into a single tree to obtain a higher prediction accuracy [29].

The general performance of RF follows the next steps,

- Given a dataset M_1, of size $m \times n$, a new dataset A_2 is created from the original data, sampling and eliminating a third part of the row data.
- The model is trained generating a new dataset through the reduced samples, estimating the unbiased error.
- At each node point (which are the points where the trees are growing simultaneously), the column n_1 is selected from the total n columns.
- When the trees finish growing, a final prediction based on the individual decisions is calculated, looking for the best classification accuracy.

For the implementation of the RF algorithm, it is used the R language [30] with the default settings of the *randomForest* library [31].

2.6. Validation

Finally, to evaluate the effectiveness of the classification process, a statistical validation is applied, based on nine metrics: true positive (TP) (conditions correctly classified), true negative (TN) (controls correctly classified), false positive (FP) (controls incorrectly classified), false negative (FN) (conditions incorrectly classified), sensitivity, specificity, positive predictive value (PPV), negative predictive value (NPV) and accuracy.

Sensitivity can be calculated by Equation (4),

$$Sensitivity = \frac{TP}{TP + FN}, \tag{4}$$

describing the true positive rate, i.e. the probability that a depressive episode is classified rightly.

Specificity can be calculated by Equation (5),

$$Specificity = \frac{TN}{FN + TP}, \tag{5}$$

describing the true negative rate, i.e. the probability that a non-depressive episode is classified rightly.

The PPV value can be defined by Equation (6), being the probability that a new episode of a person suffering from depression is classified as a depressive episode

$$PPV = \frac{Sensitivity * Prevalence}{Sensitivity * Prevalence + (1 - Specificity) * (1 - Prevalence)}, \tag{6}$$

where *Prevalence* is the percentage of observations with a condition, in this case depressive episodes.

The NPV value can be defined by Equation (7), being the probability of a episode with absence of depression is classified as negative.

$$NPV = \frac{Specificity * (1 - Prevalence)}{(1 - Sensitivity) * Prevalence + Specificity * (1 - Prevalence)}, \tag{7}$$

Finally, the accuracy can be calculated with Equation (8), being the total probability that one episode is classified correctly.

$$Accuracy = \frac{TruePositive + TrueNegative}{TruePositive + TrueNegative + FalsePositive + FalseNegative}, \tag{8}$$

3. Experiments and Results

In Figure 2 is presented a comparison of the motor activity between a control and a condition in different hours of the day. In the differences can be observed that every hour activity of control and condition shows different patterns. From these data, a segmentation is applied to form data intervals containing the information of one hour time lapses. The structure of the data for every observation is contained by 61 columns; one column for the monitored hour and one column for each minute (60 columns) of motor activity. This segmentation allowed the classification of depressive episodes per hour.

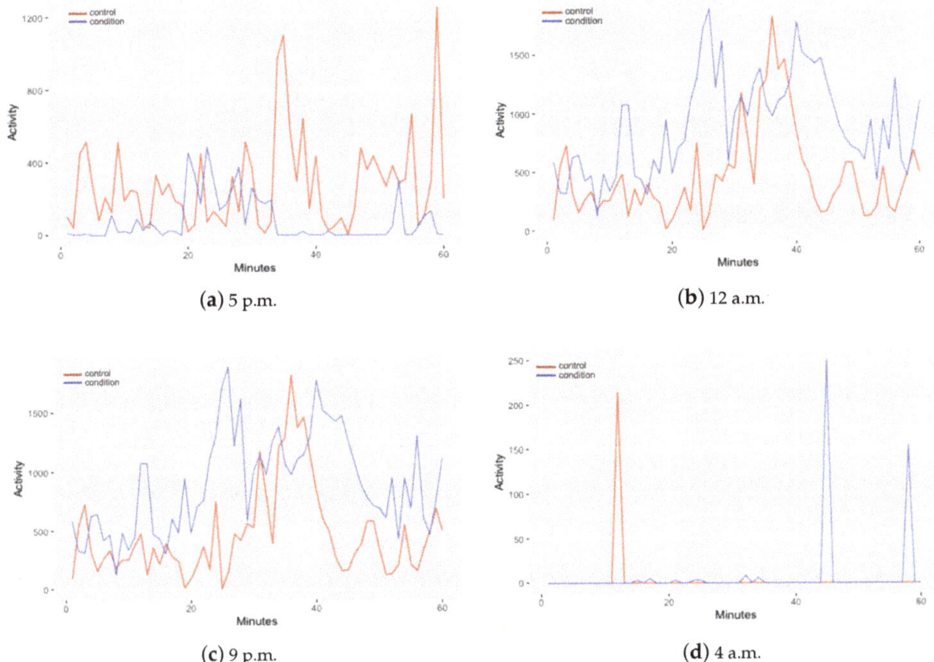

Figure 2. Comparison of motor activity in different hours of the day between a control an a condition.

From Figure 2 they can also be distinguished different patterns on the activity of control and condition subjects in different moments of the day. In Figure 2a, at 5 p.m., the control subject signal collected presents higher levels of activity in contrast with the depressive patient. This can be the most expected conduct of a patient with depression. Nevertheless, as the day ends the signal of both, control and condition, starts to change as shown in Figure 2b–d, depressive activity containing higher values.

Therefore, based on this, the data is treated in three different sets, each one corresponding to different moments of the day. One set corresponding the day, one to the night and one to the full day.

Then, for the next stage a feature selection is proposed. The results for each dataset are shown in Table 4. Accuracy is the metric used to evaluate the performance of the models constructed by the FS approach including different number of features (two, four, five, six, seven, eight, nine and ten).

In case of Night Data and Full Day Data datasets, higher accuracy is achieved with nine-features model in classification of depressive and non-depressive episodes . Day data best model is comprised by 8 features, however, the difference with nine-features model is less than 0.1 percent.

Table 4. Forward Selection results for the night, day and full day data.

Night Data		Day Data		Full Day Data	
Features	**Accuracy**	**Features**	**Accuracy**	**Features**	**Accuracy**
[2,13]	0.7398	[2,13]	0.7525	[2,13]	0.7393
[2, 12, 13]	0.7683	[2, 12, 13]	0.7589	[2,12,13]	0.7623
[2,12,13,23]	0.7706	[0,2,12,13]	0.7628	[0,2,12,13]	0.7655
[2,8,12,13,23]	0.7763	[2,8,11,22,23]	0.7728	[0,2,12,13,15]	0.7725
[2,6,8,12,13,23]	0.7776	[0,2,9,12,13,23]	0.7728	[0,2,9,12,13,15]	0.7747
[2,6,8,12,13,15,23]	0.7792	[1,2,7,8,11,22,23]	0.7728	[0,2,9,12,13,15]	0.7747
[0,2,6,8,12,13,15,23]	0.7811	[1,2,5,7,8,11,22,23]	0.7744	[0,2,5,9,12,13,15,23]	0.7758
[0,2,6,7,8,12,13,15,23]	0.7818	[0,1,2,7,9,12,13,15,23]	0.7735	[0,2,5,7,9,12,13,15,23]	0.7792
[0,2,6,7,8,9,12,13,15,23]	0.7818	[0,1,2,5,7,9,12,13,15,23]	0.7731	[0,2,5,6,7,9,12,13,15,23]	0.7761

After the feature selection with FS approach, validation step is done in two steps. Firstly, classification is done with the best nine-features model for each dataset, mentioned in Table 5, even when Day Data has higher accuracy with 8 features, best nine features is selected to compare the performance in same circunstances with the other two datasets. The results of this classification are shown in Table 6, described as Best 9 Features Day, Best 9 Features Full Day and Best 9 Features Night. In addition to accuracy, which was used in FS step, sensitivity and specificity were calculated in this validation to give a wide view of the performance of the models.

Secondly, classification is performed using the nine-features set of the Best 9 Features Night, applied to the Day Data and the Full Day Data datasets, because this nine features model is the one which outperforms all other nine features models in all proposed metrics. This in order to evaluate the performance of a general model, i.e. a unique model for all the time of the day. The results of this classification are described as the Best Model Day, Best Model Full Day in Table 6.

Table 5. Best nine features model in every dataset; day, night and full day.

Dataset	Best Nine Features
Day	kurtosis (time), mean (time), median (time), minimum (time), trim mean (time), median (frequency), SD (frequency), coefficient of variance (frequency), spectral flatness (frequency)
Night	kurtosis (time), median (time), interquartil rank (time), minimum (time), maximum (time), median (frequency), SD (frequency), coefficient of variance (frequency), spectral flatness (frequency)
Full Day	kurtosis (time), median (time), coefficient of variance (time), minimum (time), trim mean (time), median (frequency), SD (frequency), coefficient of variance (frequency), spectral flatness (frequency)

From this table can be observed that every model has a significant performance in the classification of depressive episodes. Sensitivity values oscillate from 98.24% to 99.37% and specificity range oscillates between 98.08% and 99.31%, establishing an almost perfect classification of depressive and non-depressive episodes.

The lowest, but still being good results, are those from the Day Data with nine features selected from the Night Data.

Table 6. RF results for datasets using the nine features models and the best model from the forward selection on night dataset.

Dataset	TP	TN	FP	FN	Sensitivity	Specificity	PPV	NPV	Accuracy
Best 9 Features Day	1455	2729	40	26	98.24%	98.56%	97.32%	99.06%	98.45%
Best Model Day	1470	2736	25	19	98.72%	99.09%	98.33%	99.31%	98.96%
Best 9 Features Full Day	2703	5047	53	31	98.87%	98.96%	98.08%	99.39%	98.93%
Best Model Full Day	2737	5047	19	31	98.88%	99.62%	99.31%	99.39%	99.36%
Best 9 Features Night	1259	2315	2	8	99.37%	99.91%	99.84%	99.66%	99.72%

4. Discussion

In this section the discussion of the results obtained for the different stages applied in this work is presented. Initially, the features extracted from the data are submitted to a feature selection to remove redundant or non-significant features, preserving those that contribute most to the description of depressive subjects in the different moments of the day. Then, a classification is carried out, modeling the set of features that presented the best result in the previous step for the different moments of the day. A final validation is applied to statistically evaluate the performance of the models obtained.

According to the validation values shown in the previous section, the feature selection and classification stages allowed to obtain statistically significant results.

From the feature selection step, a series of feature sets have been obtained along with their calculated accuracies, shown in Table 4. The main objective in this step is to be able to select the smallest set of features obtaining the best accuracy. As can be seen, the accuracy follows an increase pattern for most cases each time a feature is increased in the set. However, for the three data sets it is observed that the accuracy stops increasing when the tenth feature is added, and even decreases for the Day Data and Full Day Data sets. For the Night Data and the Full Day Data, the best accuracy is calculated with the set of nine features, obtaining 0.7818 and 0.7792, respectively, while for the Day Data the best accuracy is calculated with the set of eight features, obtaining 0.7744. And, in general, of all the feature sets selected for the three data sets, the best accuracy is obtained with the nine-features set of the Night Data. Based on this and in order to make a direct comparison, the sets of nine features have been selected as the best for each data set. The description of the features included in these sets are shown in Table 5.

Comparing the best nine-features sets shown in Table 5, it can be observed that only for the Night Data the maximum (time) value is selected in the FS. This may be due to sleep disorders that make patients with depression more active at night, being possible to differentiate the level of motor activity of a person who does not have this condition, since it is regularly lower.

Another important detail shown in the description of the feature sets is the frequency-related features, since for the three different data sets the same features were selected in this domain. This demonstrates the robustness and generalization in the information provided by these features, since regardless of the time of day, it is possible to identify subjects with depression presence with the levels of activity that occur.

The next step corresponds to the modeling of the set of features selected for each dataset for a classification task, based on the RF technique. For this purpose, the nine-features sets selected for each dataset were submitted to the modeling. In addition, taking into account that of all the selected feature sets, the one with the best accuracy was the nine-feature set of the Night Data, another classification was made in the Full Day Data and Day Data using this nine-feature set. To avoid confusion, this classification is labeled as Best Model Full Day and Best Model Day, for the Full Day Data and the Day Data respectively.

It is important to mention that RF was selected for the classification since it has been used to classify the motor activity of depressed subjects in other works. Zanella-Calzada et al. [22] present the classification of depressive and no depressive episodes using RF, obtaining an accuracy of 0.893, while M. Pal et al. [32] compare the performance between RF and SVM, resulting RF more efficient even with fewer parameters to make the classification.

To measure the significance of the classification, the TP, TN, FP, FN, sensitivity, specificity, PPV, NPV and accuracy metrics were measured, obtaining the results shown in Table 6. Initially, it can be seen that the FP and FN values are not significant if they are compared with the TP and TN values, taking into account that the TP and TN are the conditions and controls, respectively, correctly classified being much higher than the FP and FN values, which are the subjects incorrectly classified. An important point to note is that the lowest number of FP and FN is obtained when the classification is carried out using the set of nine features selected for Night Data applied to the three different data sets, as can be seen in the Best Model Day and Best Model Full Day. However, the lowest number of FP and FN are obtained in the Night Data set, allowing to demonstrate that the data of these nine features, specifically in this period of the day, generate values in the levels of the motor activity that allow to identify the depressive subjects, reducing the ambiguities that may be obtained in the activity levels presented by the Day Data and Full Day Data sets.

For the rest of the results it can be seen that the highest values were obtained using the Night Day feature set, since if a comparison is made of the accuracy obtained from the classification of the nine features selected from the Data Day set and the accuracy obtained from this same data set but using the nine features of the Night Data (Best Model), an increase of 0.45% can be observed. In the case of Full Day Data it is observed the same behaviour, obtaining an increase in the accuracy of 0.43% when using the set of nine features of the Night Data. For the Night Day data, the classification accuracy is almost perfect, obtaining a value of 99.72%.

Based on this, it can be noted the great contribution generated by the set of features selected for the Night Data set, but specifically with the maximum (time) feature, since it is the main difference between this set of selected features and the others two sets, contributing to have significant behavior not only in the Night Data set, but in all sets. The maximum (time) feature represents the highest value obtained from activity level and specifically at night, it allows to identify depressive subjects almost perfectly. This may be because, according to Armitage et al. [33], around 80% of patients diagnosed with MDD suffer from sleep disorders. This represents an important change in the circadian rhythm of patients suffering from depression and a healthy persons, causing depressive subjects to have more motor activity. In Figure 2 can be notice that even in a sleepy hour for both control and condition (4 a.m.) patient suffering from MDD have more disturbances than the control subject.

Finally, it should be noted that, as mentioned above, the best results are obtained using the Night Data dataset with the set of nine features specifically selected for this dataset. While the lowest results are obtained using the Day Data data set, however, these values increase if the nine features selected for Night Data are used. For the Full Day Data set, an intermediate value can be observed between the other two datasets. Therefore, based on this, it can be known that the ambiguities in the classification of depressive subjects are greater during the day than during the night. This may be due to the fact that during the day people regularly must carry out daily activities, such as work or studies, regardless of whether they suffer from depression. While at night, the presence of this condition may be more evident due to the irregularities that it can cause to sleep, while people who do not suffer from depression can have a quieter sleep and therefore much less activity levels.

5. Conclusions

The main objective of this work is to develop a set of models that allows to classify depressive and not depressive episodes in different moments of the day (day, night and full day) based on the motor activity levels of subjects. For this purpose, a series of stages are applied to the Depresjon database, which describes the activity levels of patients with presence of depression and controls.

For the feature selection stage, it is used the FS technique based on LR, in order to obtain the set of features that provide the most relevant information for data modeling, while for the classification stage, it is applied the RF technique. For the validation of the performance of these steps, a series of statistical metrics are measured.

According to the results obtained, it can be observed that from the feature selection, the best set of selected features is obtained from the data that corresponds to the night period, since the best accuracy is calculated when classifying the subjects with these features using the activity levels presented during the night. This set is contained by nine features, being maximum (time) the feature that generates the greatest contribution, since it provides the maximum values of activity level during the night, which are generally related to the subjects that present depression.

Therefore, this allows us to conclude that it is possible to identify subjects with the presence of depression based on the model developed in this work using the data of motor activity levels. In addition, for the identification of this condition, it is sufficient for patients to only measure their activity levels through the actigraph during the night, since with these data, classification can be made through the model obtained allowing to know if the subject presents depression or not, with an accuracy of 99.72%.

It should be noted that this is a preliminary tool that can be of great support for specialists in the diagnosis of depression based on a non-invasive method, since it would only be necessary to have the patient's activity level data to make the diagnosis at through this model.

Author Contributions: Conceptualization, J.G.R.-R. and C.E.G.-T.; Data curation, J.G.R.-R., C.E.G.-T., L.A.Z.-C. and J.M.C.-P.; Formal analysis, J.G.R.-R. and C.E.G.-T.; Funding acquisition, J.I.G.-T., H.G.-R. and H.L.-G.; Investigation, J.G.R.-R., C.E.G.-T. and L.A.Z.-C.; Methodology, C.E.G.-T., L.A.Z.-C., J.M.C.-P., J.I.G.-T., H.G.-R. and M.A.S.-M.; Project administration, J.G.R.-R., C.E.G.-T., J.I.G.-T. and H.G.-R.; Resources, L.A.Z.-C., J.I.G.-T. and H.L.-G.; Software, J.M.C.-P. and R.M.-Q.; Supervision, C.E.G.-T., L.A.Z.-C., J.M.C.-P. and H.G.-R.; Validation, R.M.-Q.; Visualization, L.A.Z.-C., J.M.C.-P. and H.L.-G.; Writing—original draft, J.G.R.-R., C.E.G.-T., L.A.Z.-C. and M.A.S.-M.; Writing—review & editing, J.G.R.-R., C.E.G.-T., L.A.Z.-C., R.M.-Q. and M.A.S.-M. All authors have read and agreed to the published version of the manuscript.

Conflicts of Interest: The authors declare no conflict of interest.

References

1. Bonato, P. Advances in wearable technology and applications in physical medicine and rehabilitation. *J. Neuro Eng. Rehabil.* **2005**. [CrossRef] [PubMed]
2. Farahani, B.; Firouzi, F.; Chang, V.; Badaroglu, M.; Constant, N.; Mankodiya, K. Towards fog-driven IoT eHealth: Promises and challenges of IoT in medicine and healthcare. *Future Gener. Comput. Syst.* **2018**, *78*, 659–676. [CrossRef]
3. Garcia-Ceja, E.; Riegler, M.; Jakobsen, P.; Tørresen, J.; Nordgreen, T.; Oedegaard, K.J.; Fasmer, O.B. Depresjon: a motor activity database of depression episodes in unipolar and bipolar patients. In Proceedings of the 9th ACM Multimedia Systems Conference, ACM, Amsterdam, The Netherlands, 12–15 June 2018; pp. 472–477.
4. Cheniaux, E.; Silva, R.d.A.d.; Santana, C.M.; Filgueiras, A. Changes in energy and motor activity: Core symptoms of bipolar mania and depression? *Braz. J. Psychiatry* **2018**, *40*, 233–237. [CrossRef] [PubMed]
5. World Health Organization. *Depression and Other Common Mental Disorders: Global Health Estimates*; Technical Report; World Health Organization: Geneva, Switzerland, 2017.
6. de Psiquiatría, A.A. *Manual Diagnóstico y Estadístico de los Trastornos Mentales*, 5th ed.; DSM-5; Asociación Americana de Psiquiatría: Arlington, VA, USA, 2014.
7. Unützer, J.; Katon, W.; Callahan, C.M.; Williams, J.W., Jr.; Hunkeler, E.; Harpole, L.; Hoffing, M.; Della Penna, R.D.; Noel, P.H.; Lin, E.H.; et al. Depression treatment in a sample of 1801 depressed older adults in primary care. *J. Am. Geriatr. Soc.* **2003**, *51*, 505–514. [CrossRef]
8. Saeb, S.; Zhang, M.; Kwasny, M.; Karr, C.J.; Kording, K.; Mohr, D.C. The relationship between clinical, momentary, and sensor-based assessment of depression. In Proceedings of the 2015 IEEE 9th International Conference on Pervasive Computing Technologies for Healthcare (PervasiveHealth), Istanbul, Turkey, 21–23 May 2015; pp. 229–232.
9. Cohen, Z.D.; DeRubeis, R.J. Treatment selection in depression. *Annu. Rev. Clin. Psychol.* **2018**, *14*. doi:10.1146/annurev-clinpsy-050817-084746. [CrossRef]
10. Hamilton, M. A rating scale for depression. *J. Neurol. Neurosurg. Psychiatry* **1960**, *23*, 56. [CrossRef]

11. Montgomery, S.A.; Åsberg, M. A new depression scale designed to be sensitive to change. *Br. J. Psychiatry* **1979**, *134*, 382–389. [CrossRef]

12. Martin, L.; Poss, J.W.; Hirdes, J.P.; Jones, R.N.; Stones, M.J.; Fries, B.E. Predictors of a new depression diagnosis among older adults admitted to complex continuing care: implications for the depression rating scale (DRS). *Age Ageing* **2007**, *37*, 51–56. [CrossRef]

13. Sharp, R. The Hamilton rating scale for depression. *Occup. Med.* **2015**, *65*, 340. [CrossRef]

14. Worboys, M. The Hamilton Rating Scale for Depression: The making of a "gold standard" and the unmaking of a chronic illness, 1960–1980. *Chronic Illn.* **2013**, *9*, 202–219. [CrossRef]

15. Berle, J.O.; Hauge, E.R.; Oedegaard, K.J.; Holsten, F.; Fasmer, O.B. Actigraphic registration of motor activity reveals a more structured behavioural pattern in schizophrenia than in major depression. *BMC Res. Notes* **2010**, *3*, 149. [CrossRef] [PubMed]

16. Gao, S.; Calhoun, V.D.; Sui, J. Machine learning in major depression: From classification to treatment outcome prediction. *CNS Neurosci. Ther.* **2018**, *24*, 1037–1052. [CrossRef] [PubMed]

17. Mumtaz, W.; Ali, S.S.A.; Yasin, M.A.M.; Malik, A.S. A machine learning framework involving EEG-based functional connectivity to diagnose major depressive disorder (MDD). *Med Biol. Eng. Comput.* **2018**, *56*, 233–246. [CrossRef] [PubMed]

18. Deo, R.C. Machine learning in medicine. *Circulation* **2015**, *132*, 1920–1930. [CrossRef] [PubMed]

19. Huber, D.L.; Thomas, D.G.; Danduran, M.; Meier, T.B.; McCrea, M.A.; Nelson, L.D. Quantifying activity levels after sport-related concussion using actigraph and mobile (mHealth) technologies. *J. Athl. Train.* **2019**, *54*, 929–938. [CrossRef] [PubMed]

20. Srinivasan, R.; Chen, C.; Cook, D. Activity recognition using actigraph sensor. In Proceedings of the Fourth International Workshop on Knowledge Discovery form Sensor Data (ACM SensorKDD'10), Washington, DC, USA, 25 July 2010; pp. 25–28.

21. Acebo, C.; Sadeh, A.; Seifer, R.; Tzischinsky, O.; Wolfson, A.R.; Hafer, A.; Carskadon, M.A. Estimating sleep patterns with activity monitoring in children and adolescents: how many nights are necessary for reliable measures? *Sleep* **1999**, *22*, 95–103. [CrossRef]

22. Zanella-Calzada, L.A.; Galván-Tejada, C.E.; Chávez-Lamas, N.M.; Gracia-Cortés, M.; Magallanes-Quintanar, R.; Celaya-Padilla, J.M.; Galván-Tejada, J.I.; Gamboa-Rosales, H. Feature Extraction in Motor Activity Signal: Towards a Depression Episodes Detection in Unipolar and Bipolar Patients. *Diagnostics* **2019**, *9*, 8. [CrossRef]

23. Ravi, N.; Dandekar, N.; Mysore, P.; Littman, M.L. Activity recognition from accelerometer data. *AAAI* **2005**, *5*, 1541–1546.

24. Galván-Tejada, C.; García-Vázquez, J.; Brena, R. Magnetic field feature extraction and selection for indoor location estimation. *Sensors* **2014**, *14*, 11001–11015. [CrossRef]

25. Foster, K.R.; Koprowski, R.; Skufca, J.D. Machine learning, medical diagnosis, and biomedical engineering research-commentary. *Biomed. Eng. Online* **2014**, *13*, 94. [CrossRef]

26. Thanh Noi, P.; Kappas, M. Comparison of random forest, k-nearest neighbor, and support vector machine classifiers for land cover classification using Sentinel-2 imagery. *Sensors* **2018**, *18*, 18. [CrossRef] [PubMed]

27. Breiman, L. Random forests. *Mach. Learn.* **2001**, *45*, 5–32. [CrossRef]

28. Denisko, D.; Hoffman, M.M. Classification and interaction in random forests. *Proc. Natl. Acad. Sci. USA* **2018**, *115*, 1690–1692. [CrossRef] [PubMed]

29. Subudhi, A.; Dash, M.; Sabut, S. Automated segmentation and classification of brain stroke using expectation-maximization and random forest classifier. *Biocybern. Biomed. Eng.* **2020**, *40*, 277–289. [CrossRef]

30. R Core Team. *R: A Language and Environment for Statistical Computing*; R Foundation for Statistical Computing: Vienna, Austria, 2019.

31. Liaw, A.; Wiener, M. Classification and Regression by randomForest. *R News* **2002**, *2*, 18–22.

32. Pal, M. Random forest classifier for remote sensing classification. *Int. J. Remote Sens.* **2005**, *26*, 217–222. [CrossRef]

33. Armitage, R. Sleep and circadian rhythms in mood disorders. *Acta Psychiatr. Scand.* **2007**, *115*, 104–115. [CrossRef]

Article

Concurrent Validity and Reliability of an Inertial Measurement Unit for the Assessment of Craniocervical Range of Motion in Subjects with Cerebral Palsy

Cristina Carmona-Pérez [1,2], Juan Luis Garrido-Castro [3,4], Francisco Torres Vidal [3], Sandra Alcaraz-Clariana [2,5], Lourdes García-Luque [2], Francisco Alburquerque-Sendín [4,5,*,†] and Daiana Priscila Rodrigues-de-Souza [5,†]

1   Centro de Recuperación Neurológica de Córdoba (CEDANE), 14005 Córdoba, Spain; mcarperes@yahoo.es
2   Doctoral Program in Biomedicine, University of Córdoba, 14004 Córdoba, Spain; m72alcls@uco.es (S.A.-C.); lgarcial05@hotmail.com (L.G.-L.)
3   Department of Computer Science and Numerical Analysis, Rabanales Campus, University of Córdoba, 14071 Córdoba, Spain; cc0juanl@uco.es (J.L.G.-C.); frantorresvidal@gmail.com (F.T.V.)
4   Maimonides Biomedical Research Institute of Cordoba (IMIBIC), 14004 Córdoba, Spain
5   Physiotherapy Section, Faculty of Medicine and Nursing, University of Córdoba, 14004 Córdoba, Spain; drodrigues@uco.es
*   Correspondence: falburquerque@uco.es
†   These authors contributed equally to this manuscript.

Received: 29 December 2019; Accepted: 30 January 2020; Published: 1 February 2020

Abstract: Objective: This study aimed to determine the validity and reliability of Inertial Measurement Units (IMUs) for the assessment of craniocervical range of motion (ROM) in patients with cerebral palsy (CP). Methods: twenty-three subjects with CP and 23 controls, aged between 4 and 14 years, were evaluated on two occasions, separated by 3 to 5 days. An IMU and a Cervical Range of Motion device (CROM) were used to assess craniocervical ROM in the three spatial planes. Validity was assessed by comparing IMU and CROM data using the Pearson correlation coefficient, the paired t-test and Bland–Altman plots. Intra-day and inter-day relative reliability were determined using the Intraclass Correlation Coefficient (ICC). The Standard Error of Measurement (SEM) and the Minimum Detectable Change at a 90% confidence level (MDC_{90}) were obtained for absolute reliability. Results: High correlations were detected between methods in both groups on the sagittal and frontal planes (r > 0.9), although this was reduced in the case of the transverse plane. Bland–Altman plots indicated bias below 5°, although for the range of cervical rotation in the CP group, this was 8.2°. The distance between the limits of agreement was over 23.5° in both groups, except for the range of flexion-extension in the control group. ICCs were higher than 0.8 for both comparisons and groups, except for inter-day comparisons of rotational range in the CP group. Absolute reliability showed high variability, with most SEM below 8.5°, although with worse inter-day results, mainly in CP subjects, with the MDC_{90} of rotational range achieving more than 20°. Conclusions: IMU application is highly correlated with CROM for the assessment of craniocervical movement in CP and healthy subjects; however, both methods are not interchangeable. The IMU error of measurement can be considered clinically acceptable; however, caution should be taken when this is used as a reference measure for interventions.

Keywords: inertial sensors; pediatric neurological disease; kinematics

---

1. Introduction

Cerebral palsy (CP) comprises a group of disorders affecting the development of movement and posture, causing activity limitations, and is attributed to a non-progressive damage to the developing brain during the fetal period or in the first years of life [1]. According to the Surveillance of Cerebral Palsy in Europe, CP affects between 1 to 3 per 1000 live births [2,3], with a prevalence of 3 to 4 cases per 1000 among school-age children in the US [4]. Currently, CP is recognized as being the most common cause of serious permanent physical disability in childhood, although the prospect of survival in children with severe disability has increased in recent years. Cerebral palsy is associated with sensory deficits, cognitive deficits, communication and motor disabilities, behavioral problems, seizure disorders, pain and secondary musculoskeletal problems, with spastic paresis being one of the most common forms of presentation [5,6], affecting the magnitude of movement and motor control [7,8], including the craniocervical region. Thus, head movement alterations can impair temporomandibular joint functions [9], and increase the risk of falls [10]. Furthermore, certain disorders affecting the senses can lead to unusual head movements and these alterations of the head movements can in turn further affect the senses [11,12]. In addition, it is suggested that the evaluation of motor disorders should not be centered only on posture, but also on the analysis of movement [13]. All of the above increases the need for valid and reliable methods to study cervical movement in patients with CP.

Most of the assessment methods in CP are based on subjective measures that classify motor participation based on functional abilities [14–16]; however, more advanced approaches are necessary in clinical settings and research [17]. Inertial Measurement Units (IMUs) have been known to benefit motion assessments due to their portability, ease-of-application, and low energy consumption, in contrast to other complex electromagnetic devices or video-based optoelectronic systems, which can only be used in laboratory settings [18]. In fact, IMUs represent a scientific advancement in the bio-healthcare sector, by measuring the kinematics of body segments, since these are adapted to each body region and use specific protocols that must be validated [18–20]. Good reliability results regarding optical motion capture have been described for the assessment of cervical and thoracolumbar range of motion (ROM) [21,22]. Their use in neurological diseases includes balance assessments in multiple sclerosis [23,24], Parkinsonian tremor [25,26], or range of motion (ROM) in stroke [27]. Nevertheless, further studies are necessary to confirm the clinical and predictive importance of measurements with IMUs [13,23]. Additionally, future research is required to support this validity with other tools [28] in pediatric pathologies [18,29]. To date, in children with CP, spasticity in lower limbs has been studied, obtaining satisfactory results in terms of precision and reliability, superior to other alternatives, such as goniometry [28], and gait analysis [30].

Thus, the aim of this study was to determine the clinimetric characteristics of IMU, in terms of validity and reliability, for the assessment of cervical ROM in patients with CP. In addition, we sought to establish error threshold values and minimum detectable difference with IMU in the assessment of the cervical spine in patients with CP, in order to determine clinical effect. We hypothesized that IMU would show good concurrent validity with cervical range of motion device (CROM) and that the determination of ROM using IMU would reveal good intra- and inter-day reliability.

2. Materials and Methods

2.1. Subjects

A clinical measurement study assessing validity and reliability was designed using a two-stage repeated measures design. Patients with CP were recruited using non-probabilistic sampling of consecutive cases from the private Neurological Recovery Center of Córdoba (CEDANE) and the Rehabilitation Service of the Reina Sofía University Hospital of Córdoba (Andalusian Health Service), in Spain. The inclusion criteria were: male and female subjects aged between 4 and 14 years old; diagnosed with CP and poor head control; with the necessary cognitive and behavioral skills required for understanding tasks and following simple instructions for active participation in the study; Gross

Motor Function Classification System (GMFCS) levels I-IV; medically stable. In addition, to ensure active movement against gravity, all subjects had to achieve, at least, a level of 3 in the Manual Muscle Test of cervical muscles [31,32]. The exclusion criteria were: aggressive or self-injurious behavior; involuntary or uncontrollable movements of the head; orthopedic surgery at least 1 year before the evaluation or 6 months from the administration of botulinum toxin; anti-spasticity medications at the time of the assessment; severe tactile hypersensitivity that hinders body alignment; severe visual limitations; bone deformities or contractures that prevent assessment; history of uncontrolled pain; participation in another biomedical research (and/or patients in a period of exclusion).

Control subjects were also selected for this study. These were subjects with no neurological or other impairments, matched for gender and age (±2 years). They were recruited from the Hospital and the University, as well as via the researchers' personal contacts.

The parents or caregivers of all study subjects gave their informed consent in accordance with the tenets of the Declaration of Helsinki for inclusion before they participated in the study. The protocol was approved by the Ethics Committee of Reina Sofía University Hospital (act n°270, reference 3680, 6 November 2017 approved).

The sample size required to test the concurrent validity between the IMU and CROM was based on a bilateral Pearson's correlation coefficient, assuming an expected correlation of $r \geq 0.60$, a level of significance of 5%, and 90% power. Thus, we determined that at least 21 subjects were necessary in each group. In addition, based on previous studies [33–35], and considering an intraclass correlation coefficient (ICC) of 0.8, an accuracy of 0.23 and a level of significance of 5%, the estimated sample should comprise, at least 22 subjects (Tamaño de la muestra 1.1® software, Bogotá, Colombia). Due to the short follow-up period, no data loss was expected.

2.2. IMU Assessment

An IMU Shimmer3 ® sensor (Dublin, Ireland) was located on the patient's forehead, attached to the head using a flexible and adjustable strap (Figure 1A). Orientation in the three planes of movement was obtained by a sensor at 50 Hz, connected to an android mobile phone using iUCOTrack © (Córdoba, Spain) [21,22] a software program for the acquisition and processing of the raw data obtained by the sensor, producing kinematic results. The patient performed three movements in each of the three spatial planes (flexion and extension in the sagittal plane, left and right rotation in the transverse plane, left and right lateroflexion in the frontal plane), recording the maximum values of each movement. The ICC among the three repetitions of each movement was over 0.8 in all cases.

2.3. CROM Assessment

The Cervical Range of Motion (CROM 3 ®, Lindstrom, MN, USA) device was used for the goniometry assessment, together with the IMU. This device has three spheres (2 inclinometers and a compass) to determine the ROM in the three spatial planes (Figure 1B). Its validity and reliability have been proven in cervical functional assessments for all movements [36,37]. The CROM cannot be adapted to fit different head sizes. Thus, semi-rigid foams were used to adjust the CROM to the children's heads and to prevent any movement. As the CROM was applied together with IMU, three repetitions of each movement were also performed, for which the ICC of the three repetitions was over 0.75 in all cases.

2.4. Muscle Tone Assessment

Due to the influence of spasticity in ROM, muscle tone was assessed for flexor, extensor, and sternocleidomastoid muscles of CP subjects, using the Modified Ashworth Scale (MAS) [38,39]. This scale is widely used and easy to administrate, with moderate to good reliability in CP [40].

The MAS scale is scored as follows:

0: No increase in muscle tone.

1: Slight increase in muscle tone, manifested by a catch and release, or by minimal resistance at the end of the range of motion when the affected part(s) is moved in flexion or extension.

1+: Slight increase in muscle tone, manifested by a catch, followed by minimal resistance throughout the remainder (less than half) of the ROM.

2: More marked increase in muscle tone through most of the ROM, but affected part(s) easily moved.

3: Considerable increase in muscle tone, passive movement difficult.

4: Affected part(s) rigid in flexion or extension.

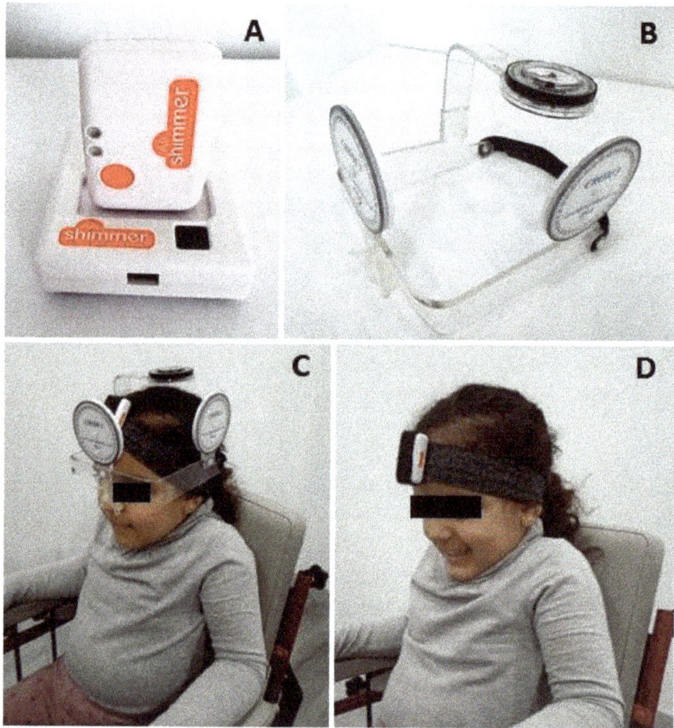

Figure 1. Devices and procedure of assessment. (**A**) Inertial Measurement Unit (IMU) Shimmer3 ® (Dublin, Ireland); (**B**) Cervical Range of Motion (CROM) 3 ® device (Lindstrom, MN, USA); (**C**) positioning of IMU and CROM to assess craniocervical range of motion (first assessment on the first day); (**D**) positioning of IMU to assess craniocervical range of motion (second assessment on the first day, and assessment on the second day).

2.5. Procedures

The general recommendations for assessments in this patient profile were applied, meaning that evaluation and treatment strategies must include relatives or caregivers who are functionally involved and part of the daily relationship (relatives/caregiver/child) [41,42].

The evaluations were performed in a quiet room, with no other people present besides the subject, assessors, and relatives/caregiver. All people stood behind the study subject, except for the assessor, who read the CROM values. A non-swivel chair was used, adapted to the anthropometric characteristics of each subject, who were seated in a standardized manner, and secured with straps when necessary. Specific instructions were given to the subject for the performance of each movement, as follows: for flexion, "first, tuck in your chin, then move your head forward and down as far as possible"; for extension, "first, raise your chin, then move your head backward, looking up as far as possible until limited by tightness or discomfort"; for rotation in each direction, "turn your head, gazing at an imaginary horizontal line on the wall, as far as possible"; for lateral flexion in each direction, "stare straight ahead and side-bend your neck by moving your ear toward your shoulder as far as possible". To avoid thoracic movement, the instructions were, "do not move your shoulders or change the amount of pressure applied to the backrest of your chair" [37]. Manual stabilization was provided during each movement to avoid movements other than those requested and to control for any proprioceptive or other sensorimotor problems that could occur during the static posture or the performance of the movements, when necessary. To control for the appearance of resistance to movement due to spasticity, an assessor performed stretches of the muscle, repositioning the joint in the position where the resistance appeared. Subsequently, a second examiner annotated the CROM values [43].

The Wong–Baker facial pain scale [44] was applied to assess whether patients suffered from pain throughout the evaluations. Its results were applied to interrupt the patient's participation in the study.

The two movements in each spatial plane were added to obtain the ROM in each plane (flexion-extension range: flexion plus extension; rotational range: right rotation plus left rotation; side-bending range: right lateral flexion plus left lateral flexion). The use of the ROM in each plane has been described as an advantage to assess cervical movement due the possible discrepancies in determining the neutral position when half movements are assessed [45].

Data were collected on two different occasions, separated 3 to 5 days. On the first day, measurements were applied twice, separated by 5 min, without changing the position of the subject. On the first evaluation, both IMU and CROM were applied, to compare results between both devices, and on the second evaluation only IMU was used, for intra-day reliability purposes. The IMU assessment was repeated 3–5 days later, to analyze inter-day reliability (Figure 1C,D).

The assessor was blinded to the previous measures at the time of the new trial [18]. All intra-day and inter-day tests were performed by the same assessor, a physiotherapist with more than 15 years of experience in the evaluation of patients with CP.

2.6. Statistical Analysis

Frequencies and percentages were used to describe categorical variables. The arithmetic mean, standard deviation and 95% confidence intervals (95% CI) were used for quantitative variables, once normality and homoscedasticity were tested (Shapiro-Wilk and Levene's tests, $p > 0.05$).

Spearman's rho correlation coefficient (r_s) was used to identify associations between cervical muscle tone and ROM, assessed with the CROM and IMU. Correlation coefficient values were considered poor when values were below 0.20, fair for values between 0.21 to 0.50, moderate from 0.51 to 0.70, very strong from 0.71 to 0.90, and almost perfect from 0.91 to 1.00 [46]. To identify possible specific characteristics in craniocervical ROM of CP, IMU, and CROM data from each assessment and spatial plane were compared between groups by unpaired t-tests.

2.6.1. Concurrent Validity Analysis

To assess concurrent validity, the Pearson's correlation coefficient (r) was applied for data obtained by IMU and CROM when applied together, that is, during the first assessment on the first day, with the same interpretation based on the Spearman rho [46]. The paired t-test was also used to analyze the differences between the means of both methods in each spatial plane ROM. In addition, Bland–Altman plots were constructed for each ROM [47,48]. The mean bias, defined as the average of the differences between both methods of measurement, was determined, together with limits of agreement (LoA), providing an estimate of the interval where 95% of the differences between both methods lie, and defined as the bias ±1.96 standard deviations of differences.

2.6.2. Reliability Analysis

The relative reliability of the measurements of each ROM evaluated with the IMU was determined by calculating ICC for intra-day and inter-day reliability (ICC2,1) [49]. The intra-day reliability was calculated based on the assessments performed on the first day, and the inter-day reliability was estimated between the first assessment on the first day and the assessment performed on the second day. For all analyzes, ICC values were considered poor when values were below 0.20, reasonable from 0.21 to 0.40, moderate from 0.41 to 0.60, good from 0.61 to 0.80, and very good from 0.81 to 1.00 [34].

The absolute reliability was determined by calculating the SEM and the Minimum Detectable Change at 90% confidence level (MDC_{90}) for each movement:

$$SEM = SD_{pooled} \times \sqrt{1 - ICC,}$$

where SD_{pooled} is the standard deviation of the scores from all subjects;

$$MDC_{90} = SEM \times \sqrt{2} \times 1.64.$$

The SEM provides a value for the random measurement error in the same unit as the measurement itself quantifies the variability within the subject and reflects the amount of measurement error for any given test (intra-day reliability) and for any test occasion (inter-day reliability) [50,51]. The MDC is an estimate of the smallest amount of change between separate measures that can be objectively detected as a true change outside the measurement error [50,52], and the MDC_{90} is frequently used to identify the effectiveness of an intervention [33].

For a better control of type I error risk, due to the repeated comparison among CROM and IMU data, a two-way ANOVA, with Evaluation (CROM; IMU first assessment on the first day; IMU second assessment on the first day; IMU assessment on the second day) as the within-subject factor, and Group (CP group; control group) as the between-subjects factor, was performed for each spatial plane ROM. The evaluation-by-group interaction and both factors were of interest. Should the interaction or any of both factors reveal significance, the Bonferroni's post-hoc test was used to verify whether a difference existed between the groups and/or within groups (view Supplementary Material, Table S1).

All hypothesis tests were bilateral and considered significant if p was less than 0.05. The data were managed and analyzed with IBM-SPSS®, version 25 (Armonk, NY, USA).

3. Results

The present study consisted of 46 participants (CP group: $n = 23$; Control group: $n = 23$), 61% of whom ($n = 28$) were female. Their average age was 8.9 years with a standard deviation of 3.2 years. The GMFCS showed that 47.8% of the CP subjects were classified as level I, 17.4% as level II, 4.4% as level III, and 30.4% as level IV. No patient showed a value of 2 or more in any muscle and over 30% of CP subjects had no impairment in muscle tone, according to the MAS. This means that muscle tone suffered, at most, a slight increase in the CP subjects. No study subjects suffered pain or other difficulties when undergoing the complete evaluation. Other basic descriptive characteristics of the groups are given in Table 1.

The correlation analysis between MAS and ROM, assessed by the CROM and the IMU, showed a common trend, with flexor, extensor, and sternocleidomastoid muscles of CP subjects significantly and negatively correlated with rotational ROM (in all cases: the higher the muscle tone, the lower the ROM). Thus, the tone of flexor and extensor muscles correlated with: CROM: $r_s = -0.504$; IMU first assessment on the first day: $r_s = -0.510$; IMU second assessment on the first day: $r_s = -0.483$; IMU assessment on the second day: $r_s = -0.412$. Right and left sternocleidomastoid muscle tone correlated with: CROM: $r_s = -0.433$; IMU first assessment on the first day: $r_s = -0.437$; IMU second assessment on the first day: $r_s = -0.420$; IMU assessment on the second day: $r_s = -0.410$. No correlation was identified in the planes of flexion-extension and side-bending.

Table 1. Demographics and clinical characteristics of the subjects.

	CP Group (*n* = 23)	Control Group (*n* = 23)	*p*-Value
Age (years)	9.2 (3.2)	8.7 (3.3)	0.594
Sex (women/men)	14/9	14/9	
Weight (kg)	28.3 (12.7)	34.6 (16.8)	0.161
Height (m)	1.32 (0.20)	1.36 (0.20)	0.503
BMI (kg/m^2)	15.5 (3.4)	17.5 (3.5)	0.049*
GMFCS level (frequency)	I: 11; II: 4; III: 1; IV: 7	-	-
Flexor muscles tone level (frequency)	0: 10; 1: 7; 1+: 6	-	-
Extensor muscles tone level (frequency)	0: 10; 1: 7; 1+: 6	-	-
Right sternocleidomastoid muscles tone level (frequency)	0: 9; 1: 8; 1+: 6	-	-
Left sternocleidomastoid muscles tone level (frequency)	0: 9; 1: 8; 1+: 6	-	-

Quantitative data are expressed as mean (standard deviation). Abbreviations: GMFCS, Gross Motor Function Classification System; BMI, body mass index. * indicates $p < 0.05$.

No differences were detected between CP subjects and controls in each ROM for any of the assessments, regardless of the method of measurement ($p > 0.05$).

Additionally, as reported in Table S1, the two-way ANOVA of the ROM of the three spatial planes showed a consistent pattern, with neither evaluation-by-group interaction nor Group factor significance, although the Evaluation factor detected statistical differences ($p \leq 0.02$). The post-hoc analysis of the Evaluation factor showed differences between CROM and the IMU assessments, with no differences among the three IMU assessments. The only exception to this pattern was the post-hoc analysis of the Evaluation factor concerning the rotational plane ROM, with statistical differences, exclusively, between the CROM and IMU assessments on the second day.

3.1. Concurrent Validity

The measurements obtained by the first IMU assessment on the first day correlated highly with the measurements of the CROM for flexion-extension and side-bending ranges in both groups (r > 0.9), although rotation range correlations were smaller (0.6 < r < 0.8). Significant differences between both methods were observed in all ROMs, with the exception of the rotational range in the control group (Table 2).

Table 2. Concurrent validity between the first IMU assessment performed on the first day of measurements and CROM by groups.

Spatial Plane	IMU First Assessment on the First Day (Standard Deviation)	CROM Assessment (Standard Deviation)	Pearson r (*p*-Value)	Student's t-test (*p*-Value)
CP group (*n* = 23)				
Flexion-Extension	133.3 (24.6)	129.0 (22.4)	0.969 (<0.001)	−3.333 (0.003)
Rotational	153.5 (19.9)	145.3 (14.7)	0.601 (0.003)	−2.396 (0.026)
Side-bending	91.3 (25.7)	96.1 (23.3)	0.916 (<0.001)	2.236 (0.036)
Control group (*n* = 23)				
Flexion-Extension	137.0 (24.5)	133.3 (23.9)	0.992 (<0.001)	−5.771 (<0.001)
Rotational	147.0 (19.0)	146.1 (16.1)	0.786 (<0.001)	−0.371 (0.714)
Side-bending	90.7 (17.5)	94.9 (18.9)	0.949 (<0.001)	3.413 (0.002)

Abbreviations: IMU, Inertial Measurement Unit; CROM, Cervical Range of Motion; CI, confidence interval; CP, cerebral palsy. Evaluation data are expressed in degrees.

The Bland–Altman plots (Figures 2 and 3) indicated bias below 5° between both measurement systems for craniocervical ROMs, except for the rotation range of the CP group (mean bias: 8.2°).

Nevertheless, the distance between LoAs for all ROMs and both groups were over 23.5°, with the exception of flexion-extension range in the control group (distance between LoAs: 12.1°). Finally, some outliers were found on the Bland–Altman plots of the CP group.

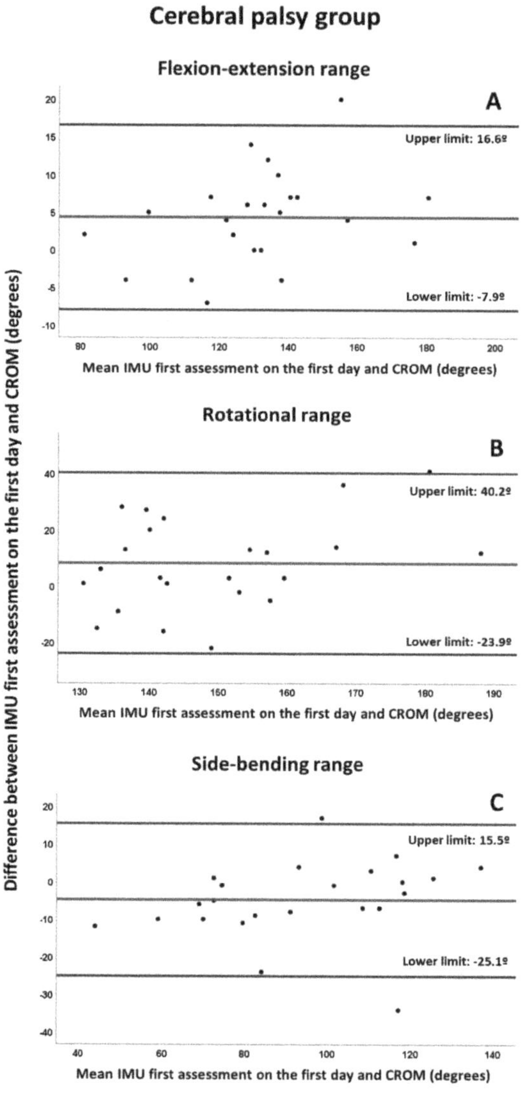

Figure 2. Bland–Altman plots for craniocervical ranges of CP group measured by IMU first assessment on the first day and the CROM in (**A**) flexion-extension range, (**B**) rotational range, (**C**) side-bending range. The red line indicates the mean bias, whereas the blue lines refers to its upper and lower limits (mean ± 1.96 standard deviation). All plots were developed as follows: the Y axis corresponds to the differences between the paired values of both methods (IMU-CROM), whereas the X axis represents the respective value of the average of both (IMU + CROM/2).

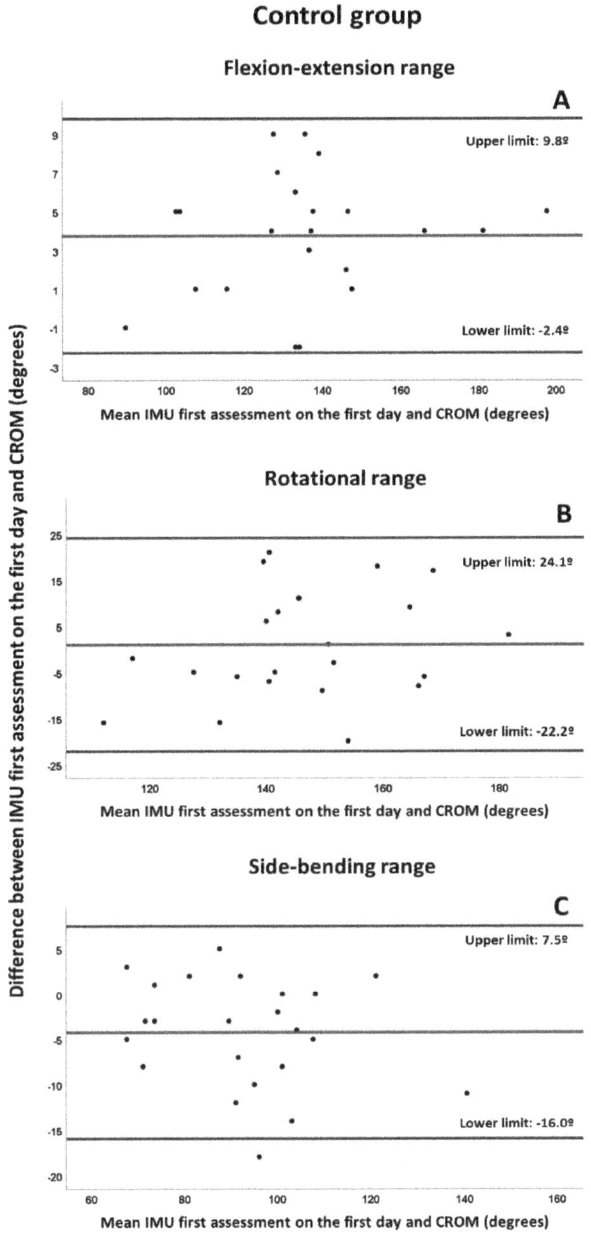

Control group

Flexion-extension range

A

Upper limit: 9.8º

Lower limit: -2.4º

Mean IMU first assessment on the first day and CROM (degrees)

Rotational range

B

Upper limit: 24.1º

Lower limit: -22.2º

Mean IMU first assessment on the first day and CROM (degrees)

Side-bending range

C

Upper limit: 7.5º

Lower limit: -16.0º

Mean IMU first assessment on the first day and CROM (degrees)

Difference between IMU first assessment on the first day and CROM (degrees)

Figure 3. Bland–Altman plots for craniocervical ranges of the Control group measured by IMU first assessment on the first day and the CROM in (**A**) flexion-extension range, (**B**) rotational range, (**C**) side-bending range. The red line indicates the mean bias, whereas the blue lines refer to its upper and lower limits (mean ± 1.96 standard deviation). All plots were developed as follows: the Y axis corresponds to the differences between the paired values of both methods (IMU-CROM), whereas the X axis represents the respective value of the average of both (IMU + CROM/2).

3.2. Intra-Day and Inter-Day Reliability

Table 3 shows absolute reliability results. For the intra-day reliability, all ICCs for both groups were from 0.82 to 0.93, with the 95% CI showing a common trend of (upper limit: ICC+0.2, lower limit: ICC−0.2). Nevertheless, absolute reliability data were variable, although all SEM were below 8,5°, and MDC_{90} between 11.4° (side-bending range of the control group), and 19.4° (rotational range of CP group).

Table 3. Intra-day and inter-day reliability of the IMU by groups.

Spatial Plane	IMU Second Assessment on the First Day (Standard Deviation)	ICC (95%CI)	SEM (°)	MDC_{90} (°)
		Intra-Day Reliability		
CP group (n = 23)				
Flexion-Extension	138.8 (26.5)	0.900 (0.762, 0.958)	8.0	18.6
Rotational	151.3 (20.1)	0.821 (0.600, 0.920)	8.3	19.4
Side-bending	91.3 (23.7)	0.925 (0.822, 0.968)	6.7	15.5
Control group (n = 23)				
Flexion-Extension	135.0 (24.7)	0.893 (0.750, 0.955)	7.9	18.4
Rotational	147.6 (20.3)	0.902 (0.750, 0.961)	5.4	12.6
Side-bending	86.3 (16.1)	0.913 (0.772, 0.965)	5.0	11.4
IMU assessment on the second day (standard deviation)		**Inter-day reliability**		
CP group (n = 23)				
Flexion-Extension	140.8 (26.4)	0.873 (0.680, 0.947)	9.0	21.1
Rotational	156.2 (19.5)	0.533 (0.117, 0.803)	13.3	30.9
Side-bending	94.4 (24.8)	0.890 (0.743, 0.953)	8.3	19.3
Control group (n = 23)				
Flexion-Extension	140.2 (29.5)	0.831 (0.602, 0.928)	11.0	25.6
Rotational	153.2 (19.1)	0.846 (0.601, 0.935)	6.5	15.1
Side-bending	93.1 (16.5)	0.864 (0.653, 0.946)	6.1	14.2

Abbreviations: IMU, Inertial Measurement Unit; ICC, Intraclass Correlation Coefficient; CI, confidence interval; SEM, Standard Error of Measurement; MDC, Minimum Detectable Change. Evaluation data are expressed in degrees.

For the inter-day reliability, all ICC values were higher than 0.8 and the 95% CI also showed a trend of (ICC+0.2; ICC−0.2), except for rotational range of the CP group (ICC = 0.53) with a wide 95% CI. SEM showed higher values compared to the intra-day values in all ROMs and both groups. Furthermore, the SEM and MDC_{90} were higher than 20° for flexion-extension and rotational ranges in CP subjects, and for flexion-extension range in controls.

4. Discussion

This is the first methodological study to assess the validity and reliability of IMUs for the assessment of craniocervical ROM in CP. SEM and MDC values were also provided for possible applications in clinical settings. According to the results, the hypotheses were partially confirmed. Thus, high correlations were found between the IMU and the CROM, although there were statistical differences when data from both methods were compared, and the range between LoAs was high. In summary, the IMU showed good concurrent validity regarding CROM; however, the methods were not interchangeable. In addition, intra-day and inter-day reliability were, in general, very good; however, the SEM and MDC were too high for inter-day comparisons for both CP and healthy subjects, hampering their application in clinical settings.

Limited information is available in the literature regarding the validity and reliability of portable devices for the measurement of cervical ROM in CP [45], which hampers comparisons with other studies. According to previous research, the validity and reliability of IMUs depend on the specific context where they are applied [18]. Thus, our results show an almost perfect correlation between the IMU and CROM, except in the case of the rotational plane, and a more questionable agreement. The exception found for the rotational plane may be due to the location of the transverse plane, where rotational movements are included, and which is perpendicular to the sagittal and frontal planes, where flexion-extension and side-bending movements are included, respectively. Flexion, extension, right lateral flexion, and left lateral flexion of the cervical spine, when originated from the neutral position, are performed in the direction of the force of gravity, making these easier to perform and, consequently, easier to reproduce in a homogenous fashion, compared to rotations, which require a continuous balance against gravity, an action which is compromised in CP [13]. Indeed, the lower validity of the results concerning the rotational plane has been consistently reported in healthy adult subjects [53], thus highlighting the significant challenge related to the measurement of ROM in rotational plane movements.

The validation pattern found in this study agrees with a former report by Chang et al. [54], who identified high correlations, although differences in craniocervical ROM values were reported, specifically for side-bending to the right and rotation to the left, when an electromagnetic portable device is compared to a universal goniometer. Although some methodological differences can be described between the research by Chang et al. and our own, such as the use of a sample comprising only healthy subjects, the assessment of movements which was performed from the neutral position, the use of a universal goniometer, and the absence of inter-day assessments, the interpretation of the agreement between both methods were the same. Thus, although the range between LoAs only defines the intervals of agreements, and not whether those limits are acceptable or not [48], we agree with Chang et al. that LoAs over 12° impaired the interchangeability of measurement methods [54]. Indeed, the assessment of craniocervical motion poses a greater challenge, compared to the motion of peripheral joints, for several reasons. First, because multiple joints are involved in craniocervical mobility. Second, due to the difficulty of avoiding thoracic spine movements, which can significantly modify the magnitude of the movements. Third, craniocervical motion is three-dimensional and movement in one axis (primary movement) can be influenced by those in other spatial planes (coupled movements) [54], which can vary among individuals [55] and can be altered by the presence of diseases [56]. All these circumstances may have influenced our results, as CP is commonly associated with a loss of cervical motor control [11,12].

Analogic goniometry using CROM has been previously established as the method of reference for evaluating neck motion [36,37], however, recently, other 3-D kinematic devices have also been proposed [57], and some circumstances support the use of digital devices to assess neck motion, such as the need of one less assessor to obtain the CROM data, a proper adjustment to the shape and size of the head without the need for additional elements (i.e., semi-rigid foams), for use in the pediatric population, and the elimination of the reading error associated with analogic devices [54,58]. This is in agreement with Paulis et al. [27], who support the objective and automatized collection of IMUs data to assess ROM in elbow muscle spasticity after stroke. Furthermore, a recent systematic review has suggested that rehabilitation research and health care services could benefit from IMUs because they provide valid data to assess ROM and joint orientation [53].

Our study showed good to very good relative reliability for intra-day and inter-day comparisons and no differences among IMU assessments in each group. It is known that the ICC increases with larger between-subjects variance [52]. In fact, we found a high variability of the data, with standard deviations over 15° in almost all cases, which means approximately 20% of the mean values of some ROMs, independent of the clinical condition. It has been described that cervical ROM shows an important dependence on age [45], which could explain the variability of the results. These interpretations of ICCs are in consonance with previous studies using IMUs in neurological diseases [27]. The exception to the high ICCs was the inter-day reliability of rotational range in the CP group, as occurred with the validity assessment. Again, it may be more difficult to repeatedly reproduce cervical rotations in a homogenous manner, as opposed to other movements, due to the balance deficiencies of subjects with CP [13]. As commented, for validity purposes, the poorer results of the rotational plane have been also found in healthy adult subjects, including ICC values below 0.8 [53]. Further research and innovative assessment approaches are necessary to improve the quality of rotational plane ROM measurements.

The SEM and MDC were acceptable for intra-day reliability, although greater for inter-day reliability, which makes their clinical applicability difficult. Thus, it is difficult to achieve an effect of more than 20° when a therapeutic intervention is applied in research or clinical settings, at least, for comparisons between different days. The previously commented high variability of the data may explain this low absolute reliability, which has been previously identified for walking performance and physical activity in CP and healthy subjects [59]. Indeed, although most studies show that the measurement error of the IMUs for motion assessment is between the 2° and 5° [18,54], the SEMs of this study were all over 5°, which can be considered clinically acceptable, according to previous studies, both in neurological patients [27] and healthy subjects [53], at least for intra-day comparisons. Furthermore, specifically for inter-day calculations, two more sources of variability may explain these results. First, spasticity varies from one movement to another, and even more when the assessments are performed on different days [60], making it difficult to ensure that the evaluations of the CP subjects were performed in the same clinical conditions. Furthermore, it is known that spasticity can be influenced by apprehension, excitement, and the position in which the child is assessed [61], which can increase the variability of the ROM results, mainly in inter-day evaluations. Second, the magnitude of a training effect or compensation cannot be calculated due to the repetitions [54], however, all the mean ROMs on the second day were higher than those of the first day in both groups, which may have also influenced the absolute reliability between days. The possible changes affecting the exact placement of the IMU between the two assessment days may also partially explain the worse absolute reliability for the inter-day comparisons [27]. Finally, although the variability is supposed to be small, calibration may be necessary for each evaluation to ensure the proper function of the gyroscope and magnetometer [54]. Previous research has identified that calibration in certain specific populations may be more challenging, such as CP patients [53].

No pain was experienced by the study subjects during the procedures, and no assessment was interrupted due to the evaluation protocol. This means that the application of IMUs for craniocervical ROM assessment is tolerable, safe, and innocuous when applied to CP and healthy children. Although the body mass index (BMI) showed differences between groups, we believe that this does not influence the study results, due to the simplicity of the task performed. Furthermore, CP subjects revealed increased BMI values compared to healthy subjects [62,63], which is a common health problem in this population.

Surprisingly, no differences in ROM were detected between CP subjects and controls, although most CP clinical presentations are associated with spasticity. However, the level of increased cervical muscle tone in the study sample can be considered as being low, which is a plausible explanation of these results. Indeed, regardless of the method of measurement, only fair to moderate correlations were found, exclusively between the tone of cervical muscles and rotational ROM, perhaps due to the fact that greater motor control is necessary to perform rotations, as previously described, with no correlations in any other spatial plane. Furthermore, the association between spasticity, hypertonia, and ROM is not completely understood at this time [60]. On the contrary, the reduction of craniocervical ROM has been described as a characteristic of several musculoskeletal and neurological diseases. Thus, cervicogenic headache in children determines reduced flexion, extension, and lateroflexion, although not rotational movements [64], plagiocephaly limits cervical ROM, especially in the rotational plane [65], and congenital muscular torticollis reduces ROM in frontal and transversal planes [66]. In conclusion, specific craniocervical ROM is not a characteristic of CP in children, at least when muscle tone is slightly increased.

It has been suggested that subjects with motor disorders could benefit from IMUs for the following three purposes: (1) Objective quantification of motor disorders; (2) Proprioceptive enhancement through visual-motor feedback; (3) Functional compensation via an inertial person-machine interface [29]. From a clinical assessment point of view for CP, IMUs have been successfully applied for the stimulation and analysis of activity using interactive games [67], for the assessment of lower limb spasticity [58], during gait [30], and for the assessment of specific characteristics in the cervical spine in small samples [11,68]. Following the increasing interest and evidence of the benefits of IMU application in pathological populations, in terms of guiding clinical decision making (e.g., quantify deficits and determine progress in time) [69], the current study adds the assessment of cervical ROM to the field of research of IMU in CP.

Despite the promising results of the current study, some limitations were identified. The applicability of findings is limited to similar samples and assessment protocols. A wider scope is necessary to establish conclusions regarding specific GMFCS levels or other age ranges. Furthermore, the current study only assessed the ROM of simple movements in a specific and controlled setting, which limits the applicability of the results to more complex tasks and day-to-day conditions [18]. In fact, although more simple movements are used to produce better clinimetric properties [70], this approach did not solve the common measurement problems of rotational plane mobility [53]. The sample size was relatively small, and several variables showed a high variability, which could have affected the strength of the comparisons. No inter-assessor reliability was evaluated, although the automatized process with IMUs makes an inter-assessor error difficult, as commented. Finally, some previous research has recommended the use of two IMUs to assess cervical ROM [45,71], but we preferred the application of one IMU adding a manual stabilization during each movement to avoid unwanted body motions, due to the difficulties to maintain a thoracic sensor fixed in children, and the need of an additional support on trunk in some CP subjects. Further research is necessary, considering additional factors, such as other movement characteristics, including velocity, acceleration or coupled angles, and innovative assessment protocols, with a special focus on complex and day-to-day tasks and rotational plane movements, and larger sample sizes, in order to standardize technical procedures and obtain accurate and normative data [53].

5. Conclusions

A high correlation was found between IMU and CROM for the assessment of craniocervical motion among individuals with CP and healthy subjects. However, both methods are not interchangeable. In CP subjects, the error of measurement in IMU can be considered clinically acceptable for the sagittal and frontal planes, although not for the transverse plane. When used as a reference measure for interventions, neck ROMs must achieve very high changes to ensure that the detected changes are significant. Future studies should be conducted to establish the normative data of craniocervical ROMs for specific population subgroups.

Supplementary Materials: Supplementary materials can be found at http://www.mdpi.com/2075-4418/10/2/80/s1.

Author Contributions: Conceptualization, C.C.-P., J.L.G.-C., F.A.-S., and D.P.R.-d.-S.; methodology, C.C.-P., J.L.G.-C., F.T.V., F.A.-S., and D.P.R.-d.-S.; formal analysis, J.L.G.-C. and F.A.-S.; investigation, C.C.-P., F.T.V., S.A.-C., D.P.R.-d.-S., and L.G.-L.; writing—original draft preparation, C.C.-P., J.L.G.-C., F.T.V., S.A.-C., F.A.-S., D.P.R.-d.-S., and L.G.-L.; writing—review and editing, C.C.-P., J.L.G.-C., F.A.-S., and D.P.R.-d.-S.; project administration and funding acquisition, J.L.G.-C., F.A.-S., and D.P.R.-S. All authors have read and agree to the published version of the manuscript.

Funding: This research was funded by Consejería de Salud (Andalusian Government, Spain), grants PI-0324-2017 and PIN-0079-2016. The funding sponsor had no role in the design of the study; in the collection, analyses, or interpretation of data; in the writing of the manuscript, and in the decision to publish the results.

Acknowledgments: To the staff of the Rehabilitation Service of the Reina Sofía University Hospital of Córdoba (Andalusian Health Service, Spain) for their support during the recruitment of the sample and the data collection.

Conflicts of Interest: The authors declare no conflict of interest.

References

1. Bax, M.; Frcp, D.M.; Rosenbaum, P.; Dan, B.; Universitaire, H.; Fabiola, R.; De Bruxelles, U.L.; Goldstein, M.; Pt, D.D.; Rosenbaum, P. Review Proposed definition and classification of cerebral palsy. *Dev. Med. Child Neurol.* **2005**, *47*, 571–576. [CrossRef] [PubMed]
2. Odding, E.; Roebroeck, M.E.; Stam, H.J. The epidemiology of cerebral palsy: Incidence, impairments and risk factors. *Disabil. Rehabil.* **2006**, *28*, 183–191. [CrossRef] [PubMed]
3. Zeitlin, J.; Mohangoo, A.; Delnord, M. *European Perinatal Health Report. The health and care of pregnant women and babies in Europe in 2010*, 1st ed.; Euro-Peristat: Paris, France, 2013; pp. 182–189.
4. Winter, S.; Autry, A.; Boyle, C.; Yeargin-Allsopp, M. Trends in the prevalence of cerebral palsy in a population-based study. *Pediatrics* **2002**, *110*, 1220–1225. [CrossRef] [PubMed]
5. Blair, E. Epidemiology of the cerebral palsies. *Orthop. Clin. North Am.* **2010**, *41*, 441–455. [CrossRef] [PubMed]
6. Agarwal, A.; Verma, I. Cerebral palsy in children: An overview. *J. Clin. Orthop. Trauma* **2012**, *3*, 77–81. [CrossRef]
7. Papageorgiou, E.; Simon-Martinez, C.; Molenaers, G.; Ortibus, E.; Van Campenhout, A.; Desloovere, K. Are spasticity, weakness, selectivity, and passive range of motion related to gait deviations in children with spastic cerebral palsy? A statistical parametric mapping study. *PLoS ONE* **2019**, *14*, e0223363. [CrossRef]
8. McDowell, B.C.; Salazar-Torres, J.J.; Kerr, C.; Cosgrove, A.P. Passive range of motion in a population-based sample of children with spastic cerebral palsy who walk. *Phys. Occup. Ther. Pediatr.* **2012**, *32*, 139–150. [CrossRef]
9. Redstone, F.; West, J.F. The importance of postural control for feeding. *Pediatr. Nurs.* **2004**, *30*, 97–100.
10. Holt, K.G.; Ratcliffe, R.; Jeng, S. Head stability in walking in children with cerebral palsy and in children and adults without neurological impairment. *Phys. Ther.* **1999**, *79*, 1153–1162. [CrossRef]
11. Velasco, M.A.; Raya, R.; Muzzioli, L.; Morelli, D.; Otero, A.; Iosa, M.; Cincotti, F.; Rocon, E. Evaluation of cervical posture improvement of children with cerebral palsy after physical therapy based on head movements and serious videogames. *Biomed Eng Online* **2017**, *16*, S74. [CrossRef]
12. Gresty, M.A.; Halmagyi, G.M. Abnormal head movement. J. Neurol. Neurosurg. *Psychiatry* **1979**, *42*, 705–714.
13. Velasco, M.A.; Raya, R.; Ceres, R.; Clemotte, A.; Ruiz Bedia, A.; Gonzalez Franco, T.; Rocon, E. Positive and negative motor signs of head motion in cerebral palsy: Assessment of impairment and task performance. *IEEE Syst. J.* **2016**, *10*, 967–973. [CrossRef]
14. Palisano, R.; Rosenbaum, P.; Walter, S.; Russell, D.; Wood, E.; Galuppi, B. Development and reliability of a system to classify gross motor function in children with cerebral palsy. *Dev Med Child Neurol* **1997**, *39*, 214–223. [CrossRef] [PubMed]
15. Palisano, R.J.; Rosenbaum, P.; Bartlett, D.; Livingston, M.H. Content validity of the expanded and revised Gross Motor Function Classification System. *Dev. Med. Child Neurol.* **2008**, *50*, 744–750. [CrossRef]
16. McDowell, B. The gross motor function classification system - Expanded and revised. *Dev. Med. Child Neurol.* **2008**, *50*, 725. [CrossRef]
17. Ronen, G.M.; Fayed, N.; Rosenbaum, P.L. Outcomes in pediatric neurology: A review of conceptual issues and recommendationsThe 2010 Ronnie Mac Keith Lecture. *Dev. Med. Child Neurol.* **2011**, *53*, 305–312. [CrossRef]
18. Cuesta-Vargas, A.I.; Galán-Mercant, A.; Williams, J.M. The use of inertial sensors system for human motion analysis. *Phys Ther Rev* **2010**, *15*, 462–473. [CrossRef]
19. Kim, M.; Kim, B.H.; Jo, S. Quantitative evaluation of a low-cost noninvasive hybrid interface based on EEG and eye movement. *IEEE Trans Neural Syst Rehabil Eng* **2015**, *23*, 159–168. [CrossRef]
20. Carcreff, L.; Gerber, C.N.; Paraschiv-Ionescu, A.; De Coulon, G.; Newman, C.J.; Armand, S.; Aminian, K. What is the best configuration of wearable sensors to measure spatiotemporal gait parameters in children with cerebral palsy? *Sensors (Switzerland)* **2018**, *18*, 394. [CrossRef]
21. Li, X.; González Navas, C.; Garrido-Castro, J.L. Reliability and validity of cervical mobility analysis measurement using an inertial sensor in patients with axial spondyloarthritis. *Rehabilitacion* **2017**, *51*, 17–21. [CrossRef]
22. Aranda Valera, I.C.; Mata Perdigón, F.J.; Martínez Sánchez, I.; González Navas, C.; Collantes Estévez, E.; Garrido Castro, J.L. Use of inertial sensors for the assessment of spinal mobility in axial spondyloarthritis patients. *Rehabilitacion* **2018**, *52*, 100–106. [CrossRef]

23. Solomon, A.J.; Jacobs, J.V.; Lomond, K.V.; Henry, S.M. Detection of postural sway abnormalities by wireless inertial sensors in minimally disabled patients with multiple sclerosis: A case-control study. *J. Neuroeng. Rehabil.* **2015**, *12*, 74. [CrossRef] [PubMed]

24. Spain, R.I.; Mancini, M.; Horak, F.B.; Bourdette, D. Body-worn sensors capture variability, but not decline, of gait and balance measures in multiple sclerosis over 18 months. *Gait Posture* **2014**, *39*, 958–964. [CrossRef] [PubMed]

25. Mancini, M.; Salarian, A.; Carlson-Kuhta, P.; Zampieri, C.; King, L.; Chiari, L.; Horak, F.B. ISway: A sensitive, valid and reliable measure of postural control. *J. Neuroeng. Rehabil.* **2012**, *9*, 59. [CrossRef] [PubMed]

26. Delrobaei, M.; Memar, S.; Pieterman, M.; Stratton, T.W.; McIsaac, K.; Jog, M. Towards remote monitoring of Parkinson's disease tremor using wearable motion capture systems. *J. Neurol. Sci.* **2018**, *384*, 38–45. [CrossRef] [PubMed]

27. Paulis, W.D.; Horemans, H.L.D.; Brouwer, B.S.; Stam, H.J. Excellent test-retest and inter-rater reliability for Tardieu Scale measurements with inertial sensors in elbow flexors of stroke patients. *Gait Posture* **2011**, *33*, 185–189. [CrossRef] [PubMed]

28. Van den Noort, J.; Harlaar, J.; Scholtes Rehab Med, V. Inertial sensing improves clinical spasticity assessment. *Gait Posture* **2008**, *28S*, S47. [CrossRef]

29. Raya, R.; Rocon, E.; Ceres, R.; Harlaar, J.; Geytenbeek, J. Characterizing head motor disorders to create novel interfaces for people with cerebral palsy: Creating an alternative communication channel by head motion. *IEEE Int. Conf. Rehabil. Robot.* **2011**, *2011*, 5975409.

30. Van Den Noort, J.C.; Ferrari, A.; Cutti, A.G.; Becher, J.G.; Harlaar, J. Gait analysis in children with cerebral palsy via inertial and magnetic sensors. *Med. Biol. Eng. Comput.* **2013**, *51*, 377–386. [CrossRef]

31. Hislop, H.; Avers, D.; Brown, M. *Muscle Testing: Techniques of Manual Examination and Performance Testing*, 9th ed.; Elsevier Inc.: St. Louis, MI, USA, 2014; pp. 1–6.

32. Manikowska, F.; Chen, B.P.J.; Jówiak, M.; Lebiedowska, M.K. Validation of Manual Muscle Testing (MMT) in children and adolescents with cerebral palsy. *NeuroRehabilitation* **2018**, *42*, 1–7. [CrossRef]

33. Haik, M.N.; Alburquerque-Sendín, F.; Camargo, P.R. Reliability and minimal detectable change of 3-dimensional scapular orientation in individuals with and without shoulder impingement. *J. Orthop. Sports Phys. Ther.* **2014**, *44*, 341–349. [CrossRef] [PubMed]

34. Shrout, P.E.; Fleiss, J.L. Intraclass correlations: Uses in assessing rater reliability. *Psychol. Bull.* **1979**, *86*, 420–428. [CrossRef]

35. Nakagawa, T.H.; Moriya, É.T.U.; Maciel, C.D.; Serrão, F.V. Test-retest reliability of three-dimensional kinematics using an electromagnetic tracking system during single-leg squat and stepping maneuver. *Gait Posture* **2014**, *39*, 141–146. [CrossRef] [PubMed]

36. Audette, I.; Dumas, J.P.; Côté, J.N.; De Serres, S.J. Validity and between-day reliability of the cervical range of motion (CROM) device. *J. Orthop. Sports Phys. Ther.* **2010**, *40*, 318–323. [CrossRef] [PubMed]

37. Fletcher, J.P.; Bandy, W.D. Intrarater reliability of CROM measurement of cervical spine active range of motion in persons with and without neck pain. *J. Orthop. Sports Phys. Ther.* **2008**, *38*, 640–645. [CrossRef] [PubMed]

38. Ashworth, B. Preliminary trial of carisoprodol in multiple sclerosis. *Practitioner* **1964**, *192*, 540–542. [PubMed]

39. Bohannon, R.W.; Smith, M.B. Interrater reliability of a modified ashworth scale of muscle spasticity. *Class. Pap. Orthop.* **1987**, *67*, 206–207. [CrossRef] [PubMed]

40. Numanoğlu, A.; Günel, M.K. Intraobserver reliability of modified Ashworth scale and modified Tardieu scale in the assessment of spasticity in children with cerebral palsy. *Acta Orthop. Traumatol. Turc.* **2012**, *46*, 196–200. [CrossRef]

41. Bartlett, D.J.; Palisano, R.J. Factors influencing the acquisition of motor abilities of children with cerebral palsy: Implications for clinical reasoning. *Phys Ther* **2002**, *82*, 237–248. [CrossRef]

42. Bartlett, D.J.; Palisano, R.J. A multivariate model of determinants of motor change for children with cerebral palsy. *Phys Ther.* **2000**, *80*, 598–614. [CrossRef]

43. Van Den Noort, J.C.; Scholtes, V.A.; Harlaar, J. Evaluation of clinical spasticity assessment in Cerebral palsy using inertial sensors. *Gait Posture* **2009**, *30*, 138–143. [CrossRef] [PubMed]

44. Bieri, D.; Reeve, R.A.; Champion, D.; Addicoat, L.; Ziegler, J.B. The Faces Pain Scale for the self-assessment of the severity of pain experienced by children: Development, initial validation, and preliminary investigation for ratio scale properties. *Pain* **1990**, *41*, 139–150. [CrossRef]

45. Raya, R.; Garcia-Carmona, R.; Sanchez, C.; Urendes, E.; Ramirez, O.; Martin, A.; Otero, A. An inexpensive and easy to use cervical range of motion measurement solution using inertial sensors. *Sensors (Switzerland)* **2018**, *18*, 2582. [CrossRef] [PubMed]

46. Akoglu, H. User's guide to correlation coefficients. *Turk J Emerg Med.* **2018**, *18*, 91–93. [CrossRef]

47. Bland, J.M.; Altman, D.G. Statistical methods for assessing agreement between two methods of clinical measurement. *Lancet* **1986**, *1*, 307–310. [CrossRef]

48. Giavarina, D. Understanding Bland Altman analysis. *Biochem. Medica* **2015**, *25*, 141–151. [CrossRef]

49. Koo, T.K.; Li, M.Y. A Guideline of selecting and reporting Intraclass correlation coefficients for reliability research. *J. Chiropr. Med.* **2016**, *15*, 155–163. [CrossRef]

50. Donoghue, D.; Murphy, A.; Jennings, A.; McAuliffe, A.; O'Neil, S.; Charthaigh, E.N.; Griffin, E.; Gilhooly, L.; Lyons, M.; Galvin, R.; et al. How much change is true change? The minimum detectable change of the Berg Balance Scale in elderly people. *J. Rehabil. Med.* **2009**, *41*, 343–346. [CrossRef]

51. Lexell, J.E.; Downham, D.Y. How to assess the reliability of measurements in rehabilitation. *Am. J. Phys. Med. Rehabil.* **2005**, *84*, 719–723. [CrossRef]

52. Weir, J.P. Quantifying test-retest reliability using the intraclass correlation coefficient and the SEM. *J. Strength Cond. Res.* **2005**, *19*, 231–240.

53. Poitras, I.; Dupuis, F.; Bielmann, M.; Campeau-Lecours, A.; Mercier, C.; Bouyer, L.J.; Roy, J.S. Validity and reliability ofwearable sensors for joint angle estimation: A systematic review. *Sensors (Switzerland)* **2019**, *19*, 1555. [CrossRef]

54. Chang, K.V.; Wu, W.T.; Chen, M.C.; Chiu, Y.C.; Han, D.S.; Chen, C.C. Smartphone application with virtual reality goggles for the reliable and valid measurement of active craniocervical range of motion. *Diagnostics* **2019**, *9*, 71. [CrossRef]

55. Malmström, E.M.; Karlberg, M.; Fransson, P.A.; Melander, A.; Magnusson, M. Primary and coupled cervical movements: The effect of age, gender, and body mass index. A 3-dimensional movement analysis of a population without symptoms of neck disorders. *Spine (Phila. Pa. 1976)*. **2006**, *31*, 44–50. [CrossRef]

56. Guo, L.Y.; Lee, S.Y.; Lin, C.F.; Yang, C.H.; Hou, Y.Y.; Wu, W.L.; Lin, H.T. Three-dimensional characteristics of neck movements in subjects with mechanical neck disorder. *J. Back Musculoskelet. Rehabil.* **2012**, *25*, 47–53. [CrossRef]

57. Song, H.; Zhai, X.; Gao, Z.; Lu, T.; Tian, Q.; Li, H.; He, X. Reliability and validity of a Coda Motion 3-D Analysis system for measuring cervical range of motion in healthy subjects. *J. Electromyogr. Kinesiol.* **2018**, *38*, 56–66. [CrossRef]

58. Choi, S.; Shin, Y.B.; Kim, S.Y.; Kim, J. A novel sensor-based assessment of lower limb spasticity in children with cerebral palsy. *J. Neuroeng. Rehabil.* **2018**, *15*, 45. [CrossRef]

59. Gerber, C.N.; Carcreff, L.; Paraschiv-Ionescu, A.; Armand, S.; Newman, C.J. Reliability of single-day walking performance and physical activity measures using inertial sensors in children with cerebral palsy. *Ann. Phys. Rehabil. Med.* **2019**, (in press). [CrossRef]

60. Bar-On, L.; Molenaers, G.; Aertbeliën, E.; Van Campenhout, A.; Feys, H.; Nuttin, B.; Desloovere, K. Spasticity and its contribution to hypertonia in cerebral palsy. *Biomed Res. Int.* **2015**, *2015*, 317047. [CrossRef]

61. Sarathy, K.; Doshi, C.; Aroojis, A. Clinical examination of children with cerebral palsy. *Indian, J. Orthop.* **2019**, *53*, 35–44.

62. Pascoe, J.; Thomason, P.; Graham, H.K.; Reddihough, D.; Sabin, M.A. Body mass index in ambulatory children with cerebral palsy: A cohort study. *J. Paediatr. Child Health* **2016**, *52*, 417–421. [CrossRef]

63. Rimmer, J.H.; Yamaki, K.; Lowry, D.M.D.; Wang, E.; Vogel, L.C. Obesity and obesity-related secondary conditions in adolescents with intellectual/developmental disabilities. *J. Intellect. Disabil. Res.* **2010**, *54*, 787–794. [CrossRef]

64. Budelmann, K.; Von Piekartz, H.; Hall, T. Is there a difference in head posture and cervical spine movement in children with and without pediatric headache? *Eur. J. Pediatr.* **2013**, *172*, 1349–1356. [CrossRef]

65. Murgia, M.; Venditto, T.; Paoloni, M.; Hodo, B.; Alcuri, R.; Bernetti, A.; Santilli, V.; Mangone, M. Assessing the cervical range of motion in infants with positional plagiocephaly. *J. Craniofac. Surg.* **2016**, *27*, 1060–1064. [CrossRef]

66. Lee, J.Y.; Koh, S.E.; Lee, I.S.; Jung, H.; Lee, J.; Kang, J.-I.; Bang, H. The cervical range of motion as a factor affecting outcome in patients with congenital muscular torticollis. *Ann. Rehabil. Med.* **2013**, *37*, 183–190. [CrossRef]

67. Rihar, A.; Sgandurra, G.; Beani, E.; Cecchi, F.; Pašič, J.; Cioni, G.; Dario, P.; Mihelj, M.; Munih, M. CareToy: Stimulation and Assessment of Preterm Infant's Activity Using a Novel Sensorized System. *Ann. Biomed. Eng.* **2016**, *44*, 3593–3605. [CrossRef]

68. Saiz, B.M.; Maris, E.; Mussin, P.; Lopesino, R.A.; Martı, I.; Callejo, S.H.; Lo, R.R.; Lara, S.L. Short-term effects of an intervention program through an inertial sensor (ENLAZA) for the improving of head control in children with cerebral palsy. *Gait Posture* **2016**, *49*, S227.

69. Porciuncula, F.; Roto, A.V.; Kumar, D.; Davis, I.; Roy, S.; Walsh, C.J.; Awad, L.N. Wearable Movement Sensors for Rehabilitation: A Focused Review of Technological and Clinical Advances. *PM R* **2018**, *10*, S220–S232. [CrossRef]

70. Walmsley, C.P.; Williams, S.A.; Grisbrook, T.; Elliott, C.; Imms, C.; Campbell, A. Measurement of Upper Limb Range of Motion Using Wearable Sensors: A Systematic Review. *Sport. Med. Open* **2018**, *4*, 53. [CrossRef]

71. Theobald, P.S.; Jones, M.D.; Williams, J.M. Do inertial sensors represent a viable method to reliably measure cervical spine range of motion? *Man. Ther.* **2012**, *17*, 92–96. [CrossRef]

Article

Assessment of the Status of Patients with Parkinson's Disease Using Neural Networks and Mobile Phone Sensors

Yulia Shichkina [1,*], Elizaveta Stanevich [1] and Yulia Irishina [2]

[1] St. Petersburg State Electrotechnical University "LETI", 197376 St. Petersburg, Russia; liza_soft_1313@mail.ru

[2] N.P. Bechtereva Institute of the Human Brain of the Russian Academy of Sciences, 197376 St. Petersburg, Russia; irishina@mail.ru

* Correspondence: strange.y@mail.ru

Received: 22 March 2020; Accepted: 11 April 2020; Published: 12 April 2020

Abstract: Parkinson's disease (PD) is one of the most common chronic neurological diseases and one of the significant causes of disability for middle-aged and elderly people. Monitoring the patient's condition and its compliance is the key to the success of the correction of the main clinical manifestations of PD, including the almost inevitable modification of the clinical picture of the disease against the background of prolonged dopaminergic therapy. In this article, we proposed an approach to assessing the condition of patients with PD using deep recurrent neural networks, trained on data measured using mobile phones. The data was received in two modes: background (data from the phone's sensors) and interactive (data directly entered by the user). For the classification of the patient's condition, we built various models of the neural network. Testing of these models showed that the most efficient was a recurrent network with two layers. The results of the experiment show that with a sufficient amount of the training sample, it is possible to build a neural network that determines the condition of the patient according to the data from the mobile phone sensors with a high probability.

Keywords: Parkinson's disease; recurrent neural network; smartphone; motion sensor; monitoring the condition of patients

---

1. Introduction

Parkinson's disease (PD) is one of the most common neurodegenerative disorders, only second after Alzheimer's disease. It is a chronic progressive neurodegenerative disease, associated mainly with dopamine deficiency in the subcortical ganglia of the brain and manifests primarily by movement disorders in the form of Parkinson's syndrome (hypokinesia, or slow movement combined with increased muscle tone of the limbs and trunk, or rigidity and/or tremor), as well as a wide range of non-motor manifestations. A feature of the development of PD is that the neurodegenerative process begins 10 years or more before the appearance of motor disorders, but the diagnosis of PD is possible only with the appearance of the latter. There are no laboratory or instrumental methods that can confirm the diagnosis of PD. Due to neurotransmitter replacement therapy for PD, after a few years, in most cases, the clinical picture of motor disorders is modified, which is manifested by various fluctuations in motor symptoms such as a change in the severity of symptoms during the day depending on dopaminergic therapy, as well as various violent movements (dyskinesias). Due to the progression of the disease (a decrease in the number of dopamine-producing neurons), treatment regimens in the late stages of the disease become more complex and fluctuations and dyskinesias become more pronounced [1,2].

Modern means of observing patients are limited, requiring the time of a doctor and a patient, which limits the frequency of clinical evaluations. Another way to monitor the patient's condition is to constantly fill a patient diaries. However, the information in diaries is subjective, which leads to significant changes in indicators depending on the mental state of the patient. Consequently, the results of the analysis based on diaries have low accuracy. For example, a study [3,4] showed that only in 11% of cases the patient's perception corresponds to clinical indicators.

Mobile phones provide automatic, convenient monitoring and recording of observations in real time, which can be invaluable for large-scale studies and personal monitoring of patient health. Patients can simply download the application to their smartphone, which allows the system to collect and analyze data. Developed specialized tests on mobile phones can give medical staff access to long-term measurements of the severity of symptoms and their variations.

The analysis of data obtained using smartphones is extremely difficult due to the large number of diverse types of data selected over long periods of time. The main unresolved tasks for today are: how to simultaneously analyze a wide range of symptoms associated with PD; how to best aggregate and analyze huge volumes of clinically relevant data; in what form should the analysis results be displayed. To solve such problems, it is possible to use neural networks. Research results [5] show that the use of machine learning methods can give a level of confidence in assessing a patient's condition comparable to a doctor.

This article describes our studies to assess the condition of a patient with Parkinson's disease based on data from mobile phones on the nature of the use of the phone, such as the angle of rotation and tilt of the phone. The article is organized as follows: the second section provides an overview of existing solutions in the field of monitoring the status of patients with Parkinson's disease and the distinctive features of our study; the third section provides a brief description of the data collection system and the statement of the problem; the fourth section is devoted to the analysis of possible neural network architectures used to classify the status of patients with Parkinson's disease and to describe the learning outcomes of these networks; the fifth section describes further research; the sixth and seventh sections discuss the limitations in this study and the study conclusions, respectively.

2. Overview of Related Research

The analysis of studies on the use of neural networks to monitor the status of patients with Parkinson's disease has shown that conditionally all studies can be divided into the following classes:

1. The use of neural networks for predicting Parkinson's disease in relation to individual parameters of the patient's condition. These include studies [6–9]. These studies describe various models of neural networks, as well as interesting and useful results on the simultaneous use of various types of neural networks. The main drawback of these studies is the use of neural networks only of single parameters, such as voice or tremor of limbs or others.

2. The integrated use of artificial intelligence methods for the diagnosis of Parkinson's disease. For example, studies have been conducted on the use of: A probabilistic neural network (PNN) and classification tree (CIT) [10]; Gaussian models, principal component analysis methods, linear discriminant analysis, least squares support vector method (LS-SVM), probabilistic neural network (PNN), and common neural network regression (GRNN) [11]; methods based on k-clustering medium (KMCFW) and the complex-valued artificial neural network (CVANN) [12]; combination minimum redundancy maximum relevance (mRMR) attribute selection algorithm and CVANN [13]; neural networks and machine learning methods [14]; the wavelet transforms and neural networks [15]; or radial basis function neural network (RBFNN), based on the particle swarm optimization (PSO) and principal component analysis (PCA) [16,17].

3. Comparisons between the effectiveness of the applications of artificial intelligence methods for the diagnosis of Parkinson's disease, for example, incremental search (IS), Monte Carlo search (MCS), and hybrid search (HS) [18].

It is also possible to classify existing studies by a set of parameters regarding the status of patients with Parkinson's disease, for which Parkinson's disease was diagnosed or predicted by:

4. tremor of human limbs [10];
5. voice [12,13,18];
6. genetic factors [14];
7. motor activity [15,19];
8. the results of medical devices, for example, electroencephalogram signals [9].

In all studies, the following equipment was mainly used for data collection:

9. the use of special sensors [15,19,20];
10. the use of mobile phones [21,22];
11. the use of special medical equipment [9,23].

It should be noted that most often, special sensors are used in research; for example, a set of accelerometers and gyroscopes are placed on the body of subjects when they perform a series of standard motor tasks [20]. However, such an approach to measuring the parameters of the condition of patients cannot be applied for everyday monitoring. In contrast to the authors of [15,19,20], we use tools that are more accessible to many patients to measure hand movements. However, in this case we cannot take into account the correlation between the accelerometers on the arm, trunk, and leg. Using the phone, we can, however, collect other important parameters about the condition of patients, such as evaluating memory, attention, voice parameters, emotional state, and others. This is another distinguishing feature of our research, i.e., assessment of the patient's condition by a large set of parameters.

3. Statement of the Problem

Monitoring the patient's condition is the key to the success of the correction of the main clinical manifestations of PD, including the almost inevitable modification of the clinical picture of the disease against the background of prolonged dopaminergic therapy.

Modern smartphones have built in accelerometers which promise to enable quantifying minute-by-minute activity of patients (e.g., walk or sit) [24]. The quality of smartphones makes it possible to improve medical diagnostics and monitor the patient's condition.

The data used in this study was obtained using a mobile application in which patients and healthy people solve a wide range of tests that have been agreed by medical staff. When passing the tests, the system evaluates speech, hand tremors, tapping of fingers, speed, balance, and reaction time.

Examples of application windows that the patient interacts with are shown in Figure 1.

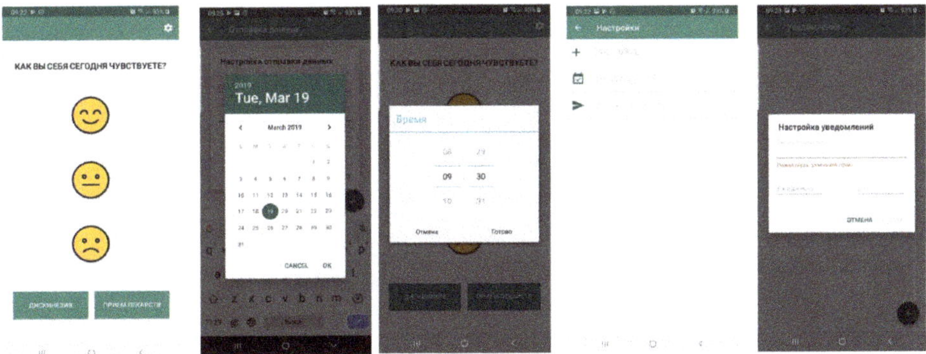

Figure 1. Examples of some patient application windows.

In previous articles [22], we described the architecture of our monitoring system for a patient with Parkinson's disease. The collected indicators are presented in Table 1. Our study involved 28 people, of which 10 patients aged 45 to 80 years with a diagnosis of PD. All subjects gave their informed consent for inclusion before they participated in the study. The study was conducted in accordance with the Declaration of Helsinki, and the protocol was approved at the meeting of the Academic Council of the N.P.Bechtereva Institute of the Human Brain of the Russian Academy of Sciences dated 17 September 2015. (No. 29). Data collection was carried out for one month. As a result, we made 100,000 records on the angles of rotation and tilt of the mobile phone and 5000 records according to the test results.

Table 1. Parameters obtained using tests on mobile phones.

Parameter	Parameter Description	Value Example	Unit
Text erased	Number of characters deleted	13	-
Text time	Time for writing text in a special test	139,763	ms
Levenstein Distance	Metric measuring the difference between two sequences of characters	5	-
Miss clicks	The number of misses when clicking buttons in the application	4	-
Miss clicks distance	The distance between the center of the nearest button and the center of the finger touch the phone screen	2.357022	dp
Azimuth	The longitudinal axis of the coordinate system	70.29761	degree
Pitch	The transverse axis of the coordinate system	−80.30805	degree
Roll	The vertical axis of the coordinate system	13.761927	degree
Tapping left count	The number of touches by the index finger of the left hand of the button on the screen in 1 min in a special test	47	-
Tapping right count	The number of touches with the index finger of the right hand of the button on the screen for 1 min in a special test	52	-
Dyskinesia	The presence of dyskinesia	1	-
Pill	The number of medications taken	4	-
State	Subjective assessment of the patient's condition (0 - poor, 0.5 - uncertain, 1 - good)	1	-
Voice volume	Voice volume	44	decibels
Voice pause	The number of pauses between words	3	-
Voice count pause	The pause time between words	4794	ms
Velocity	The speed of the phone during its active use	1.342	ms

dp or dip (density-independent pixels) is an abstract unit of measurement that allows applications to look the same on different screens and resolutions, ms is milliseconds.

To train the network, we needed data on how patients felt. Therefore, subjects entered data throughout the day after passing the tests i.e., the time of taking the medicine, the severity of dyskinesia, and assessing their condition. A numerical score was calculated after the completion of its passage according to the results of each test. Then the obtained results were analyzed together with the data entered manually. We data mined the results of the tests and data from device sensors to create a "PD score", which characterizes the severity of the disease.

The aim of the current study was to build a system for monitoring the status of a patient with Parkinson's disease based on a set of parameters that can be estimated based on data obtained using mobile phones.

To achieve this goal, it was necessary to build a number of neural networks to classify (evaluate) each of the parameters. Then, based on the obtained parameter estimates, the patient's condition was classified using a neural network. This process is shown in Figure 2.

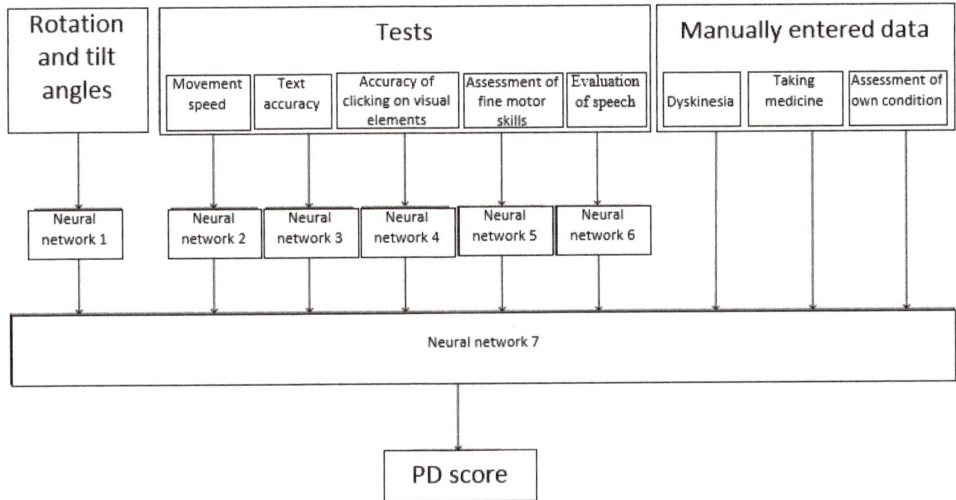

Figure 2. The process of diagnosing the severity of symptoms of Parkinson's disease (PD).

A neural network is a sequence of neurons interconnected by synapses. A neuron is a computing unit that receives information, performs simple calculations on the information, and passes the information on further. They are divided into three main types of layers: input, hidden, and output. A synapse is a connection between two neurons. Synapses have one parameter called the weight. Thanks to the synapse, the input information changes form when it is transmitted from one neuron to another. The neural network diagram is shown in Figure 3.

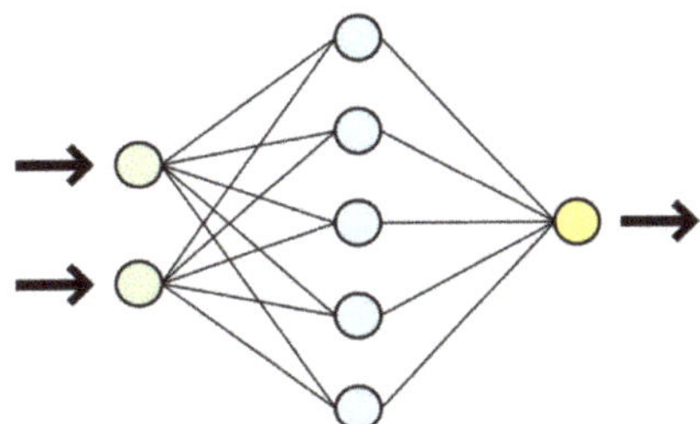

Figure 3. Scheme of a simple neural network. Green indicates input neurons, blue indicates hidden neurons, and yellow indicates output neuron.

The purpose of training a neural network is to obtain reliable results. Prediction is what the neural network returns after receiving input, for example, "given the number of drugs, the probability of tremor in a patient's hands becomes lower is 60%". Sometimes a neural network makes mistakes, but it can learn from them. If the predicted value is too high, it will reduce weight in order to get a lower predicted value next time, and vice versa.

It should be noted that in Figure 2 not all possible observable parameters are listed, however those for which data are currently being collected in the monitoring system for patients with PD are presented [22]. In the future, the range of observed parameters can be expanded, and the methods for evaluating each parameter changed.

One of the parameters used was the nature of the change in the angle of rotation and tilt of the phone while passing tests on it. Therefore, one of the tasks was used to assess changes of the angle of rotation and tilt of the mobile phone during the tests.

To solve this problem, it was necessary to:

- development of an application that allows the collection of training samples for a neural network using mobile devices;
- prepare and process data for a future neural network model;
- analyze of the possibility of using neural networks in order to classify the condition of patients with PD;
- select and test various variants of neural network architectures on the obtained sample;
- test the neural network in patients with PD;
- evaluate of the results of the neural network.

In the framework of this article, the process of evaluating changes in the values of the rotation angles and tilt of a mobile phone using a neural network is discussed in detail. For other parameters, similar studies are carried out.

4. The Choice of Neural Network Architecture, the Creation and Testing of the Network

One of the subtasks of this study was the choice of neural network architectures and the assessment of the possibility of using neural networks to classify the patient's state according to the values of rotation angles and tilt angles of the mobile phone. Therefore, in this section, various options for neural network architectures are considered, the training procedure is described, and the results of network training are analyzed.

4.1. Input and Output Data of Neural Networks within One Subtask

Fragments of the input data of the values of the rotation angles and tilt of the mobile phone for the neural network to solve the problem of the classification of the patient's state are presented in Table 2.

Table 2. An example of input data.

Azimuth	Pitch	Roll
70.29761	−80.30805	13.761927
80.7381	88.84665	−0.4941308
103.91639	89.57664	−23.674112
120.00478	89.701294	−39.76287
136.86708	89.748215	59.555275

Data in the Table 2 is sorted by azimuth. But, the azimuth values in the first and last line are not necessarily the minimum and maximum values. Such numbers were obtained in this study. In other tests, they may be different.

Obviously, each user can hold the phone in different ways and the angle of rotation and tilt of the phone will differ for different users. Therefore, to classify the condition of patients with PD, it is of interest not the absolute values of the angles, but the relative values and the frequency of their changes. The next step is to normalize the deviation values of each of the three parameters presented in Table 2 relative to the average value of the corresponding parameter. The result of the neural network is a number in the range from 0 to 1. A number is closer to 1 indicated more often the phone is in a less unbalanced state. However, the question remands whether this is it typical for patients with PD. Neural networks can help answer this question.

4.2. Neural Network Activation Functions

The activation function in neural networks determines the output signal depending on the set of input data. In hidden layers of neural networks, regardless of their architectures, the activation function is: $ReLU(x) = \max(0;x)$; in the output layer, this is the logistic activation function (sigmoid): $\sigma(x) = \frac{1}{1+e^{-x}}$. Sigmoid takes a real value as an input and displays a different value in the range from 0 to 1. It has the following properties: non-linear, continuously differentiable, monotonous, and has a fixed output range. The main disadvantage is vanishing gradients.

Unlike a sigmoid, ReLU is called a piecewise function as half of the output is linear (positive output) and the other half is non-linear. It does not suppress the neuron yield between 0 and 1, which helps with back propagation. ReLU provides the same benefits as Sigmoid, but with better performance and no vanishing gradient problem.

4.3. The Loss Function. Metrics. Neural Network Optimization Algorithm

When training a neural network, binary cross-entropy (BCE) [25] or standard deviation (MSE) can be used as a loss function:

$$BCE = -\hat{Y}\ln Y - \left(1 - \hat{Y}\right)\ln(1 - Y)$$

$$MSE = \frac{1}{n}\sum_{i=1}^{n}\left(Y_i - \hat{Y}_i\right)^2$$

The combined use of the standard deviation and the logistic activation function can lead to a situation where the gradients of the weights of the neural network turn to zero (this situation is called paralysis of the neural network), and further training becomes impossible. In this regard, binary cross-entropy is used as a loss function. The standard deviation is used as a metric of the quality of learning, since gradients are not calculated for the metrics.

When training a neural network, an optimization algorithm is implemented to minimizes the loss function. To train this neural network, the root mean square propagation (RMSProp) optimization algorithm is used [26]. This algorithm uses a moving average to normalize the gradient, which allows its output function to quickly converge to a given value.

4.4. Variants of Neural Network Architectures

As the initial data are a function of time, it is important to consider the previous values of these functions to obtain a reliable result. In this regard, to solve this problem, at least one recurrent layer must be present in the neural network, as layers of other types (fully connected and convolutional) cannot extract features from time dependencies.

Currently, two types of recurrent networks are used: Long Short-Term Memory (LSTM) and Gated Recurrent Unit (GRU).

The initial architecture of the neural network consists of four layers. The first is recurrent of 128 neurons, followed by fully connected layers of 64, 32, and one neuron. The number of neurons was selected by the heuristic method. The subsequent architectures created on the basis of the initial version will be obtained by adding layers, removing layers, and changing the type of layer without changing the number of neurons in the layer. Therefore, at the stage of training a neural network, the influence of its architecture (i.e., the number of layers and their types) on the learning outcome will be more pronounced in comparison with the influence of the number of neurons (on each layer) on the learning outcome.

As a possible improvement of the neural network, the introduction of a second recurrence layer was proposed. This increases the learning speed and accuracy of the neural network by reducing the time it takes to calculate the results of classifying the status of patients with PD based on the data provided.

This change of architecture leads to an improvement in the case if the initial data sequences form more complex dependencies that cannot be detected with a single layer.

Next, hidden fully connected layers were removed. In this case, a completely recurrent neural network was obtained, the architecture of which is shown in Figure 4. In this case, there was an improvement in the performance of computing the results (by reducing the number of layers) with only a slight deterioration in accuracy.

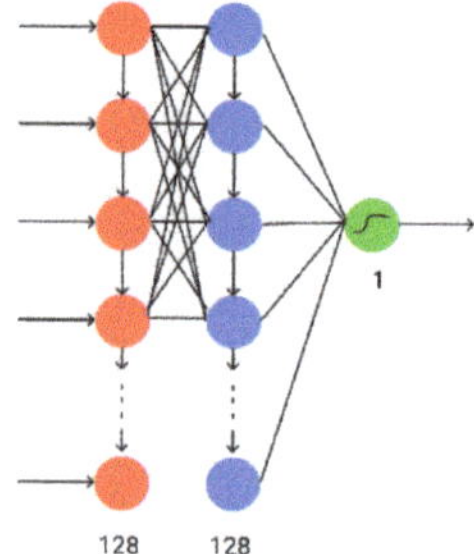

Figure 4. A fully recurrent version of a neural network. Blue and red are the first and second recurrent layers, green is the result.

For original data sets, it is advisable to choose an architecture with three hidden layers, since an advantageous compromise is achieved between system performance and accuracy.

4.5. Neural Network Training

Learning outcomes are shown in Figures 5–8. On the horizontal axis is the serial number of the training epoch, and on the vertical axis is the standard deviation of the result of neural network calculations from the corresponding value in the validation sample. The blue line indicates the loss function in the training set, and the orange line in the test set. As the returned values do not have units, the standard deviation also does not have units.

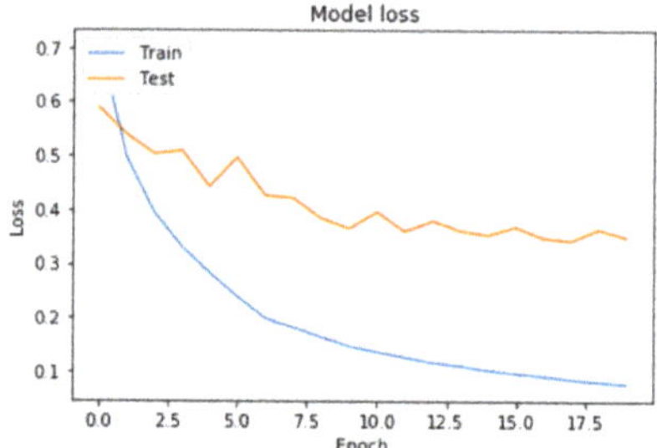

Figure 5. The results of training a neural network with one recurrent layer.

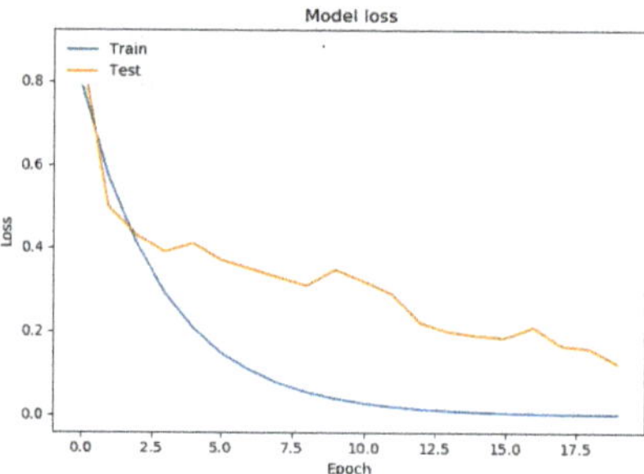

Figure 6. The results of training a neural network with two recurrent layers.

Figure 5 shows a graph of the loss function when training a neural network with one recurrent layer and two fully connected layers. It can be seen from the Figure 5 that both the learning error and the testing error monotonously decreased. This means that the neural network is capable of generalizing the source data.

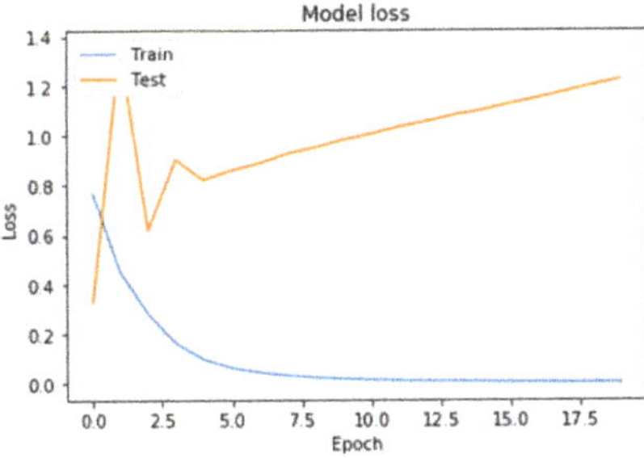

Figure 7. The results of training neural network with a fully recurrent architecture (gated recurrent unit; GRU).

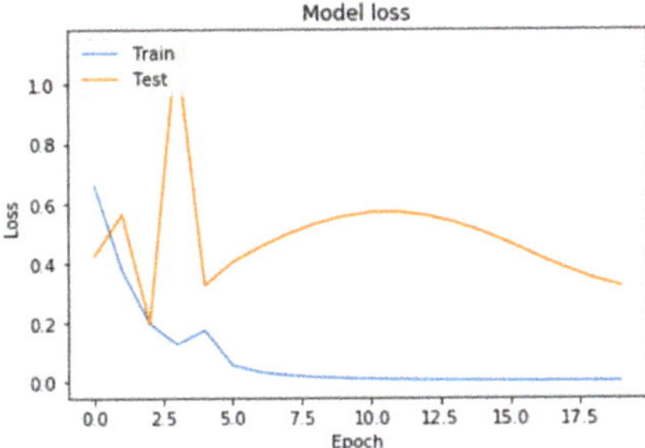

Figure 8. The results of neural network training with fully recurrent architecture (Long Short-Term Memory; LSTM).

Figure 6 shows a graph of the loss function in training a neural network with two recurrent and two fully connected layers. Learning and testing errors decreased, and the value of the testing error for this architecture was less than for a neural network with one recurrent layer, i.e., it generalized data better.

Figure 7 shows a graph of the loss function in training a neural network with two GRU recurrent layers. As a result of training a neural network using this architecture, it led to an increase in the graph of the testing error function from the beginning of training.

Figure 8 shows a graph of the loss function in training a neural network with two recurrent LSTM layers. The behavior of the test error curve did not correspond to the state of retraining. The value of the loss function was higher than that of a neural network with two recurrent layers.

Table 3 presents the results of further training of all the neural networks described above.

Table 3. The results of training of neural networks of various architectures.

Architecture	The Number of Epoch before Retraining	Loss Function (BCE)	Accuracy
1 GRU + 2 fully connected	29	0.031161	0.817
2 GRU + 2 fully connected	32	0.029795	0.88951
2 LSTM	26	0.03809	0.8528

LSTM is Data in the Long Short-Term Memory, GRU is Gated Recurrent Unit, BCE is binary cross-entropy.

An analysis of the results of training neural networks of various architectures shows that an architecture with two recurrent and two fully connected layers had greater accuracy compared to other options. Thus, a neural network with two recurrent and two fully connected layers best generalizes the data and is suitable for further training.

4.6. Neural Network Implementation

TensorFlow and Keras libraries were used to build and train neural networks. They provide implementations of various types of recurrence layers, including versions that use GPU computing. In order to train neural networks in this study, we used a back propagation method, an implementation of which is also provided in Keras.

Neural network training was implemented using the following code:

```python
# Creating a neural network of sequential execution (each layer is associated with only one subsequent)
model = Sequential()
# Creating GRU recurrence layers for 128 neurons
model.add(keras.layers.LSTM(128, return_sequences = True))
# Creating fully connected layers with 64 and 32 neurons and a ReLU activation function
model.add(Dense(64, activation = 'relu'))
model.add(Dense(32, activation = 'relu'))
# Creating a fully connected layer with 1 neuron and logistic activation function
model.add(Dense(1, activation = 'sigmoid'))
# Compilation of a neural network with binary cross-entropy as a function of losses, standard deviation as a metric
# and RMSProp gradient descent optimization algorithm
model.compile(loss = 'binary_crossentropy', optimizer = 'rmsprop', metrics = ['mse'])
# Start the training procedure, 20 epochs will be passed, 20% of the initial data are used as a validation sample
history = model.fit(X_norm, y, epochs = 20, validation_split = 0.2)
```

Using libraries significantly speeds up the development time of neural networks, allowing to quickly implement and test various architectures that eliminate the need to write low-level and template code.

5. Future Perspective

Currently, studies are being conducted on the application of the results of neural networks constructed according to tests and on changes in the rotation angles and tilt angles of the telephone to build a common neural network.

The results of a general neural network in the monitoring system for the condition of patients with PD will be used to track the dynamics of changes in the patient's condition and the effect of drugs. Further scientific research regarding the determination of the parameters that are available for measurements using mobile phones that most characterize the patient's condition will be useful.

6. Limitations

The main limitation of this work is that preliminary assessments of the conditions by the patients themselves were used for the training and evaluation of constructed neural network models. However, in the early stages of PD, data from the sensors of patients' phones cannot yet be clearly distinguishable from data on the state of a healthy person. This is a new problem for our study, and has not been solved in the world. We hope that if we can collect much more data, we will get closer to solving it. Much more data is needed both in volume and in the number of parameters in order to more accurately determine the stability of tests based on machine learning and smartphones to these mixed factors.

7. Conclusions

The availability of mobile phones to monitor patients with PD can have a profound impact on clinical practice, giving doctors access to long-term data. These additional data can help doctors obtain a more complete and objective understanding of the symptoms and fluctuations of their patients' symptoms and, therefore, will allow for more accurate diagnoses and treatment regimens.

We are creating a system of information and technological support for research on Parkinson's disease, taking into account the collection and processing of data of a large volume. Currently, in this system we have already created most of the modules, such as collecting data on the nature of the use of the phone and automatically filling out the patient's diary according to the data entered by the patient on the phone. Partially, we described this in Section 3 of this article. A more complete description can be found in one of our previous articles [22]. Using this system, data is sent daily to the doctor's computer. This eliminates the disadvantage of the lack of constant communication with the specialist, which we discussed in Section 2. The doctor has the opportunity to see a continuous graph of the dynamics of the patient's condition. We have shown that it is possible to use neural networks, which can make filling out a patient's diary without his participation and make this filling out better.

As a result of this work, it was determined that the neural network is able to summarize data on the rotation angles and tilt angles of the mobile phone, in order to classify the condition of patients with PD. In this paper, we considered and carried out a comparative analysis of various architectures for building neural networks. An architecture of a neural network with two recurrent layers was chosen, at which an acceptable level of accuracy in classifying the state of a patient with PD was achieved.

Python neural networks were built using the TensorFlow and Keras libraries. It was established that the use of LSTM blocks instead of GRU leads to greater accuracy of the neural network. Future research will focus on the construction of neural networks for more parameters and a larger sample of the source data, which will lead to the training of the neural network so that it will return even more accurate diagnosis results.

Author Contributions: Conceptualization, Y.S.; methodology, Y.S., Y.I.; software, E.S.; validation, E.S., Y.S. and Y.I.; formal analysis, E.S.; writing—original draft preparation, E.S., Y.S.; writing—review and editing, Y.S. and Y.I.; visualization, E.S.; supervision, Y.I.; project administration, Y.S.; funding acquisition, Y.S. All authors have read and agreed to the published version of the manuscript.

Funding: The research was funded by Russian Foundation for Basic Research (RFBR) according to the research project #18-57-34001.

Conflicts of Interest: The authors declare no conflict of interest.

References

1. Levin, O.S.; Fedorova, N.V. Parkinson's Disease. M.: MEDpress-inform. 2015, p. 384. Available online: http://med-press.ru/upload/iblock/c36/c368629d8d7c93da1359086510170207.pdf (accessed on 20 February 2020).

2. Illarioshkina, S.N.; Levina, O.S. *Guidelines for the Diagnosis and Treatment of Parkinson's Disease*, 2nd ed. 2017, p. 336. Available online: https://www.03book.ru/upload/iblock/871/871e24b9e93e4cc2ad5b6c7f93fde4b0.pdf (accessed on 20 February 2020).

3. Macleod, A.D.; Taylor, K.S.M.; Counsell, C.E. Mortality in Parkinson's disease: A systematic review and meta-analysis. *Mov. Disord.* **2014**, *29*, 1615–1622. [CrossRef] [PubMed]

4. Stone, A.A.; Shiffman, S.; Schwartz, J.E.; Broderick, J.E.; Hufford, M.R. Patient compliance with paper and electronic diaries. *Control. Clin. Trials* **2003**, *24*, 182–199. [CrossRef]

5. Zhang, Y.N. Can a smartphone diagnose parkinson disease? A deep neural network method and telediagnosis system implementation. *Parkinson's Dis.* **2017**, *2017*. [CrossRef] [PubMed]

6. Åström, F.; Koker, R. A parallel neural network approach to prediction of Parkinson's Disease. *Expert Syst. Appl.* **2011**, *38*, 12470–12474. [CrossRef]

7. Hirschauer, T.J.; Adeli, H.; Buford, J.A. Computer-Aided Diagnosis of Parkinson's disease using enhanced probabilistic neural network. *J. Med. Syst.* **2015**, *39*, 197. [CrossRef]

8. Ramzi, M.S.; Salah, A.M.; Abdul Rahman, K.A.; Abdul Karim, H.A.G.; Majed, R.B.; Mohamed, N.M.; Bassem, S.A.-N.; Samy, S.A.-N. Parkinson's disease prediction using artificial neural network. *Int. J. Acad. Health Med. Res.* **2019**, *3*, 1–8.

9. Oh, S.L.; Hagiwara, Y.; Raghavendra, U.; Yuvaraj, R.; Arunkumar, N.; Murugappan, M.; Acharya, U.R. A deep learning approach for Parkinson's disease diagnosis from EEG signals. *Neural. Comput. Appl.* **2018**, *12*, 056011. [CrossRef]

10. Palumbo, B.; Fravolini, M.L.; Nuvoli, S.; Spanu, A.; Paulus, K.S.; Schillaci, O.; Madeddu, G. Comparison of two neural network classifiers in the differential diagnosis of essential tremor and Parkinson's disease by 123I-FP-CIT brain SPECT. *Eur. J. Nucl. Med. Mol. Imaging* **2010**, *37*, 2146–2153. [CrossRef]

11. Hariharan, M.; Kemal, P.; Sindhu, R. A new hybrid intelligent system for accurate detection of Parkinson's disease. *Comput. Methods Programs Biomed.* **2014**, *113*, 904–913. [CrossRef]

12. Gürüler, H. A novel diagnosis system for Parkinson's disease using complex-valued artificial neural network with k-means clustering feature weighting method. *Neural Comput. Appl.* **2017**, *28*, 1657–1666. [CrossRef]

13. Peker, M.; Sen, B.; Delen, D. Computer-aided diagnosis of Parkinson's disease using complex-valued neural networks and mRMR feature selection algorithm. *J. Healthc. Eng.* **2015**, *6*, 22. [CrossRef]

14. Joshi, S.; Shenoy, D.; Vibhudendra Simha, G.G.; Rrashmi, P.L.; Venugopal, K.R.; Patnaik, L.M. Classification of Alzheimer's disease and Parkinson's disease by using machine learning and neural network methods, Second International Conference on Machine Learning and Computing. *Bangalore* **2010**, 218–222. [CrossRef]
15. Lee, S.H.; Lim, J.S. Parkinson's disease classification using gait characteristics and wavelet-based feature extraction. *Expert Syst. Appl.* **2012**, *39*, 7338–7344. [CrossRef]
16. Wu, D.; Warwick, K.; Ma, Z.; Gasson, M.N.; Burgess, J.G.; Pan, S.; Aziz, T.Z. Prediction of parkinson's disease tremor onset using a radial basis function neural network based on particle swarm optimization. *Int. J. Neural Syst.* **2010**, *20*, 109–116. [CrossRef] [PubMed]
17. Wu, D.; Warwick, K.; Ma, Z.; Burgess, J.G.; Pan, S.; Aziz, T.Z. Prediction of Parkinson's disease tremor onset using radial basis function neural networks. *Expert Syst. Appl.* **2010**, *37*, 2923–2928. [CrossRef]
18. Ene, M. Neural network-based approach to discriminate healthy people from those with Parkinson's disease, Annals of the University of Craiova. *Math. Comp. Sci. Ser.* **2008**, *35*, 112–116.
19. Keijsers, N.L.W.; Horstink, M.W.I.M.; van Hilten, J.J.; Hoff, J.I.; Gielen, C.C.A.M. Detection and assessment of the severity of Levodopa-induced dyskinesia in patients with Parkinson's disease by neural networks. *Mov. Disord.* **2000**, *15*, 1104–1111. [CrossRef]
20. Markos, G.T.; Alexandros, T.T.; Rigas, G.; Tsouli, S.; Dimitrios, I.F.; Konitsiotis, S. An automated methodology for levodopa-induced dyskinesia: Assessment based on gyroscope and accelerometer signals. *Artif. Intell. Med.* **2012**, *55*, 127–135.
21. Mark, V.A.; Toledo, S.; Shapiro, M.; Kording, K. Using mobile phones for activity recognition in Parkinson's patients. *Front. Neurol.* **2012**, *3*, 158.
22. Shichkina, Y.A.; Kataeva, G.V.; Irishina, Y.A.; Stanevich, E.V. The architecture of the system for monitoring the status in patients with parkinson's disease using mobile technologies. *Stud. Comput. Intell.* **2020**, *868*, 531–540.
23. Shiek, S.S.J.A.; Santosh, W.; Kumar, S.; Christlet, T.H.T. Neural network algorithm for the early detection of Parkinson's disease from blood plasma by FTIR micro-spectroscopy. *Vib. Spectrosc.* **2010**, *53*, 181–188.
24. Stephen, A.A.; Mark, V.A.; Konrad, P.K. Hand, belt, pocket or bag: Practical activity tracking with mobile phones. *J. Neurosci. Methods* **2014**, *231*, 22–30.
25. Mannor, S.; Peleg, D.; Rubinstein, R. The cross entropy method for classification. In Proceedings of the 22nd international conference on Machine learning, Bonn, Germany, 7–11 August 2005; pp. 561–568.
26. Tieleman, T.; LeCun, Y. Lecture 6.5-rmsprop: Divide the gradient by a running average of its recent magnitude. *COURSERA: Neural Netw. Mach. Learn.* **2012**.

Article

New Mobile Device to Measure Verticality Perception: Results in Young Subjects with Headaches

Daniel Rodríguez-Almagro, Esteban Obrero-Gaitán, Rafael Lomas-Vega *, Noelia Zagalaz-Anula, María Catalina Osuna-Pérez and Alexander Achalandabaso-Ochoa

Department of Health Science, University of Jaén, Paraje Las Lagunillas s/n, 23071 Jaén, Spain; dra00005@red.ujaen.es (D.R.-A.); eobrero@ujaen.es (E.O.-G.); nzagalaz@ujaen.es (N.Z.-A.); mcosuna@ujaen.es (M.C.O.-P.); aaochoa@ujaen.es (A.A.-O)
* Correspondence: rlomas@ujaen.es; Tel.: +34-953212918

Received: 20 August 2020; Accepted: 2 October 2020; Published: 7 October 2020

Abstract: The subjective visual vertical (SVV) test has been frequently used to measure vestibular contribution to the perception of verticality. Recently, mobile devices have been used to efficiently perform this measurement. The aim of this study was to analyze the perception of verticality in subjects with migraines and headaches. A cross-sectional study was conducted that included 28 patients with migraine, 74 with tension-type headache (TTH), and 93 healthy subjects. The SVV test was used through a new virtual reality system. The mean absolute error (MAE) of degrees deviation was also measured to qualify subjects as positive when it was greater than 2.5°. No differences in the prevalence of misperception in verticality was found among healthy subjects (31.18%), migraineurs (21.43%), or those with TTH (33.78%) ($p = 0.480$). The MAE was not significantly different between the three groups (migraine = 1.36°, TTH = 1.61°, and healthy = 1.68°) (F = 1.097, $p = 0.336$, and η2 = 0.011). The perception of verticality could not be explained by any variable usually related to headaches. No significant differences exist in the vestibular contribution to the perception of verticality between patients with headaches and healthy subjects. New tests measuring visual and somatosensory contribution should be used to analyze the link between the perception of verticality and headaches.

Keywords: mobile applications; diagnostic equipment; headache; migraine; postural balance; visual motor coordination; vestibular function tests

---

1. Introduction

Assessment of the perception of verticality is increasingly used in patients with disorders of upright body orientation [1]. It is based on a gravitational input processed in the central nervous system (CNS) from vestibular, visual, and somatosensory information [2,3] that can be measured through the use of touch, which is called haptic vertical, or by estimating the position of one's own body without the help of visual inputs, which is called subjective postural vertical (SPV) [4]. However, visual estimation of the vertical (i.e., visual vertical (VV)) is the most common test used to assess the perception of verticality in research and clinical practice [5]. The subjective visual vertical (SVV) test consists of adjusting a random-oriented line to the vertical position without the help of visual references. The initial orientation is usually between 30° and 60° right or left. The consensual values considered normal for SVV are between −2.5° and 2.5° with respect to the actual vertical [6].

It is believed that SVV tests estimate the ability of a person to perceive the gravitational vertical, and a tilt in SVV indicates vestibular imbalance in the roll plane and, thus, injuries to the utricle or its connecting nerves [7]. Although measurements of the perceived visual vertical disclose mainly vestibular dysfunctions when no cues to visual spatial orientation are provided during testing [8], several studies have found that the SVV is altered in neurological patients, mainly with stroke [9]; in subjects with spinal diseases [10]; and in patients with peripheral vestibular disorders [11,12].

An SVV test is performed classically using the bucket method, which is an easily performed and reliable bedside test for determining monocular and binocular SVV that costs less than $5 [13]. However, the bucket test is a limited method that does not allow automated data storage and is not sufficiently versatile to be able to implement different versions of VV measurements. For this reason, in recent years, various wearable methods have been created using virtual reality and mobile devices [7,14]. Most of these new methods have been tested in healthy subjects to analyze the methods' feasibility and reliability. However, experiences with subjects with health problems are scarce.

Vestibular, visual, and somatosensory systems play a major role in verticality perception [2,3]. Furthermore, it is frequently observed how disfunction in these three systems appear in conjunction with headache, taking an important part in headache development [15–17]. The possibility that patients with headaches could present some alterations in any of these three systems that may induce a misperception of VV has turned headache disorders into a study issue in relation to alteration of VV.

Headaches are a significant public health problem that affect approximately 40.5% of the global population, taking into account both migraines and tension-type headaches (TTHs) [18]. This problem is more frequent among females, university students, and urban residents [19]. Several studies have looked for an alteration of verticality in subjects with migraines and TTHs with contradictory results; some studies found no significant differences between subjects with primary headache disorders (PHD) and healthy subjects [20–22], while others found differences between healthy subjects and subjects who suffered PHD [23,24]. In view of these results, it is of interest to assess the differences in perception of VV between patients with migraine and TTH, and healthy subjects.

Additionally, some works have shown possible common factors between headache and verticality perception. A recent study of Martins et al. [25] showed a relationship between sleep disturbances and modifications in perception of verticality. Furthermore, it has been possible to observe the influence of physical activity in verticality perception [26]. In the same way, both sleep disturbances [27,28] and physical activity [29] also related to the presence of headaches and migraines. In view of the above, it might be asked whether these factors are able to explain the presence or magnitude of the alteration in perception of verticality in patients with headaches.

This study is a feasibility analysis of a new device for measuring SVV, previously validated in healthy people. The main objective of this work was to analyze the possible differences in visual perception of verticality between subjects with migraines, subjects with TTHs, and healthy subjects using a new mobile device. The secondary objective was to identify which variables usually associated with headaches could be related to SVV deviation in young students.

2. Materials and Methods

2.1. Study

To meet the objectives of this work, a cross-sectional observational study was designed, developed in accordance with the guidelines for the communication of observational studies established in the Strengthening the Reporting of Observational studies in Epidemiology (STROBE) Statement [30]. This study was carried out in accordance with the Helsinki declaration, good clinical practice, and all applicable laws and regulations and was approved by the Ethics Committee of the University of Jaén (reference number ABR. 17./7.TFM). All participants signed informed consent document.

For this study, the participants contacted us as response to posters and digital advertisements published at the University of Jaén (Jaén, Spain). The data were collected between the months of October and December 2018 at the University of Jaén. The participants had to be young subjects, university students, and older than 18 years who did not suffer from cognitive disorders; eye diseases; previous head or neck trauma; any type of acquired brain damage (ischemic or hemorrhagic stroke or damage resulting from intracranial intervention); any systemic disease with visual, vestibular, central, or musculoskeletal involvement; neuromuscular disease; or presence of neoplasia at the visual, vestibular, or central level.

2.2. Sample Size Calculation

The sample size calculation was carried out using the data obtained in the study of Asai et al. [23]. Taking into account a prevalence of migraines of approximately 20% [18] and a total prevalence of headaches of approximately 52% [31], to obtain between-group significant differences with an alpha error of 5% and a power of 80%, a minimum of 179 subjects was required.

2.3. Subjects

Two hundred and seventeen subjects were initially contacted during the month of September 2018. Of these 217 subjects, 211 were selected to participate in the present investigation after having been duly informed. Finally, 195 subjects completed all of the questionnaires and evaluations planned in the study. The selection process is graphically represented in Figure 1.

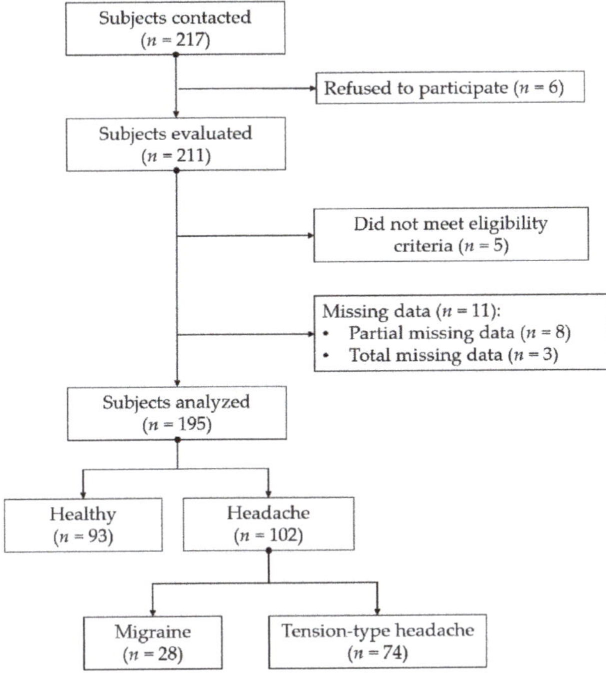

Figure 1. Flow chart of the participant selection process.

All subjects were evaluated by a physician (F.H.) who verified compliance with the eligibility criteria as well as compliance with the criteria described in the third edition of "The International Classification of Headache Disorders" [32].

2.4. Measurements

First, sociodemographic variables were recorded, including gender, age, height, weight, years at university, smoking habit, and physical activity. To quantify deviation of the perceived vertical from the theoretical vertical, the static SVV test was used through a new virtual reality system [14] during the interictal phase of headache process. The virtual reality system requires a mobile device placed into the back of a headset and a Google Cardboard-enabled application (Sistema de Realidad Virtual para Detección y Tratamiento de las Patologías Posturales y del Equilibrio, University of Jaén, Jaén, Spain, 2019) to generate a pair of stereo images (Figure 2). The test was performed in a quiet environment with dim lighting while the subject sat comfortably with their back straight and their feet uncrossed and resting on the ground. When the subject was ready, the evaluator started the test from the web application. Firstly, the virtual reality system set the line in a random position between 30° and 60° right or left. Then, the subject rotated the line using the joystick until they perceived that it was close to vertical and, then, confirmed the result with an action button. The subject had 30 second to perform each test. To calculate SVV, six measurements were made, from which the mean deviation of the perceived vertical with respect to the theoretical vertical was obtained. The mean absolute error (MAE) was calculated as the average value of the error made in each attempt, without taking the direction of deviation into account. For treatment of the dependent variable in this study, deviation of the SVV value from normal was also considered, taking as normal values those between −2.5° (left deviation) and 2.5° (right deviation) [5]. The device was validated and showed good reliability (Intraclass Correlation Coefficient (ICC) = 0.85; 95% confidence interval (CI) = 0.75–0.92) [14]. The evaluation was always made with visual correction if the patient had it prescribed.

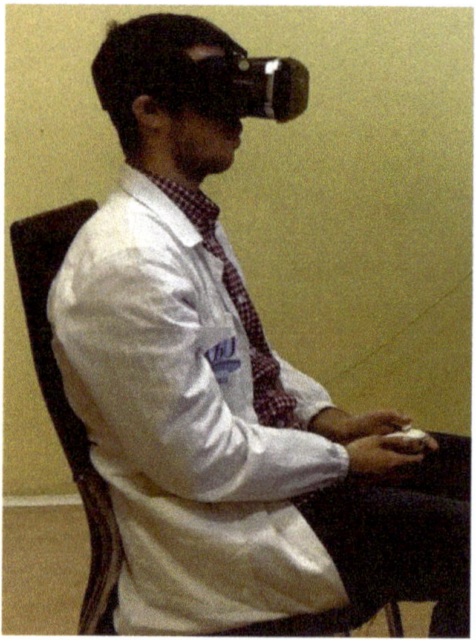

Figure 2. Participant using the mobile device to measure subjective visual vertical (SVV).

Headache-related disability as well as its frequency and intensity were assessed using the Spanish version of the migraine disability assessment (MIDAS) questionnaire [33]. This instrument is made up of seven items, the first five of which focus on three dimensions of daily life that can be affected by headaches while the remaining two items refer to the frequency and intensity of the headache. The sum of the scores of the first five items provides the degree of disability related to the headache, while the sixth and seventh items indicate the frequency and intensity of the headache, respectively. The Spanish version of the questionnaire has good reliability and validity properties [33].

The disability associated with neck pain was evaluated with the Spanish version of the "Neck Disability Index" (NDI) questionnaire [34], which is a self-administered questionnaire with 10 sections. Each of the sections offers six possible answers that represent six progressive levels of functional capacity that are scored from 0 to 5. The reliability values of this questionnaire are very high (ICC = 0.989), and it also has good internal consistency (Cronbach's α = 0.913) [34].

Sleep quality was also included as a predictor variable, due to the relationship that has been reported between the perception of verticality and the variables related to sleep [35]. To measure sleep quality, the Spanish version of the "Medical Outcomes Study Sleep Scale" (MOS-SS) was used [36], which is a self-administered questionnaire composed of 12 items, from which six subscales are extracted. From the MOS-SS questionnaire, the variables used were sleep disturbances (ICC = 0.78; 95% CI = 0.62–0.88), daytime sleepiness (ICC = 0.57; 95% CI = 0.30–0.75), sleep adequacy (ICC = 0.75; 95% CI = 0.56–0.87), snoring (ICC = 0.84; 95% CI = 0.71–0.91), waking up briefly at night due to respiratory reasons or headache (ICC = 0.84; 95% CI = 0.71–0.91), and optimal sleep (ICC = 0.76; 95% CI = 0.58–0.87), for which the reliability values were between moderate and high [36].

2.5. Statistical Analysis

Data management and analysis was carried out using the SPSS statistical package, version 23.0 (SPSS Inc, Chicago, IL, USA). The level of statistical significance was established as $p < 0.05$. The data were described using means and standard deviations for continuous variables and using frequencies and percentages for categorical variables. To determine the normality of continuous variables, the Kolmogorov–Smirnov test was used, while the Levene's test of equality of variances was used to determine the homoscedasticity of the samples.

To analyze the differences in the perception of verticality with respect to the theoretical vertical between healthy subjects, subjects with TTHs, or those with migraines, one-way analysis of variance (ANOVA) was used, while eta-squared (η2) was used to express the effect size. To evaluate differences in the prevalence of SVV alterations (SVV more than 2.5° of deviation) between subjects with TTHs or migraines and healthy subjects, the chi-square test was used.

Given the binary nature of the "alteration in the perception of verticality" variable (MAE > 2.5 or not), univariate logistic regression was used to identify which variables are related to it. The independent variables comprised sociodemographic variables; frequency, intensity, and disability associated with headaches; disability associated with neck pain; and variables related to sleep.

To identify the variables related to the degree of deviation of the perceived vertical from the theoretical vertical, univariate linear regression was used, given the continuous nature of the dependent variable. The independent variables for this analysis were the same as those used in the logistic regression.

3. Results

Of the total number of participants who completed the study, 111 were women and 84 were men. Twenty-eight subjects met the criteria for migraines, 74 met the criteria for TTHs, and 93 subjects were classified as healthy. The total prevalence of headaches in the present study was 52.3%, with 72.5% TTH and 27.5% migraines. The prevalence of verticality alterations was very similar between the three groups (Table 1). There were no statistically significant differences in the prevalence of SVV alteration ($p = 0.480$).

Table 1. Description of the samples and groups.

Categorical Variable		Migraine (n = 28)			Tension-Type Headache (n = 74)			Healthy (n = 93)			Total (n = 195)	
		F	%		F	%		F	%		F	%
Gender	Male	9	32.14		27	36.49		48	51.61		84	43.08
	Female	19	67.86		47	63.51		45	48.39		111	56.92
Smoker	Yes	3	10.71		11	14.86		10	10.75		24	12.31
	No	25	89.29		62	85.14		83	89.25		171	87.69
Physical activity	Yes	16	57.14		44	59.46		63	67.74		123	63.08
	No	12	42.86		30	40.54		30	32.26		72	36.92
School year	First	3	10.71		6	8.11		12	12.90		21	10.77
	Second	11	39.29		24	32.43		31	33.33		66	33.85
	Third	6	21.43		5	6.76		7	7.53		18	9.23
	Fourth	6	21.43		28	37.84		32	34.41		66	33.85
	Master	2	7.14		11	14.86		11	11.83		24	12.31
SVV	Yes	6	21.43	2.69 (0.46) *	25	33.78	2.87 (0.70) *	29	31.18	2.95 (0.66) *	60	30.77
>2.5	No	22	78.57	0.99 (0.47) *	49	66.22	0.96 (0.39) *	64	68.82	1.10 (0.50) *	135	69.23

Continuous Variables	Migraine (n = 28)		Tension-Type Headache (n = 74)		Healthy (n = 93)		Total (n = 195)	
	Mean	SD	Mean	SD	Mean	SD	Mean	SD
Age	20.79	2.10	21.70	4.42	21.73	3.62	21.58	3.78
Height (cm)	167.62	8.91	170.00	7.92	171.11	9.23	170.19	8.75
Weight (kg)	62.11	10.28	66.23	11.04	67.56	11.58	66.27	11.29
SVV degrees of deviation	1.36	0.84	1.61	1.04	1.68	1.02	1.61	1.01
MIDAS (Frequency of headache)	8.54	13.58	4.34	6.06	2.03	3.45	3.84	7.07
MIDAS (Intensity of headache)	5.54	1.93	5.54	2.27	2.83	2.18	4.25	2.56
MIDAS score	6.04	6.18	6.15	10.21	1.04	2.02	3.70	7.27
NDI	13.93	6.48	12.81	8.63	5.91	5.32	9.72	7.78
Sleep disturbance	41.61	15.98	40.95	13.37	36.89	15.68	39.11	14.97
Daytime somnolence	43.85	14.45	43.54	12.51	39.25	13.53	41.54	13.40
Sleep adequacy	66.37	18.49	64.19	20.07	67.38	19.34	66.03	19.46
Snoring	26.19	18.39	27.03	16.71	29.03	19.80	27.86	18.42
Awaken short	25.00	10.64	26.13	13.54	20.79	12.45	23.42	12.83
Quantity of sleep	7.43	1.23	6.91	0.89	7.09	1.16	7.07	1.08

* Mean of deviation degrees of and standard deviation for each subjective visual vertical (SVV) misperception group. MIDAS, migraine disability assessment; F, frequency; SD, standard deviation; SVV, subjective visual vertical; SF-12, 12-Item Short Form Health Survey; PCS-12, physical component summary of the SF-12; MCS-12, mental component summary of the SF-12; and NDI, neck disability index.

One-way ANOVA showed no statistically significant between-group differences in the MAE between subjects with migraines or TTHs and healthy subjects (F = 1.097, *p* = 0.336, and η2 = 0.011). The results are graphically shown in Figure 3.

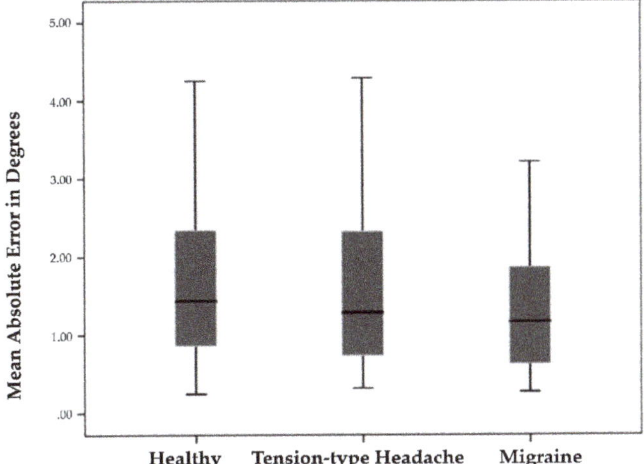

Figure 3. Between-group differences in mean absolute error in estimating subjective visual vertical (SVV).

The logistic regression performed to identify the variables related to alterations in the perception of verticality (Table 2) and the linear regression used to establish the variables that explained the degree of deviation in the MAE (Table 3) did not show statistically significant associations. An association was only found at the limit of statistical significance (*p* = 0.054) between headache-related disability and the degree of MAE deviation.

Table 2. Univariate logistic regression to analyze the factors related to alterations in the perception of verticality.

Variable	OR	95% CI		*p*-Value
		Lower	Upper	
Gender	1.087	0.533	2.219	0.818
Smoking	1.448	0.532	3.938	0.468
Physical activity	1.004	0.482	2.093	0.991
Healthy/headache	0.892	0.439	1.811	0.751
Headache Frequency	1.005	0.958	1.054	0.837
Headache intensity	1.029	0.895	1.184	0.689
MIDAS	1.023	0.981	1.067	0.280
NDI	1.014	0.970	1.060	0.535
Sleep disturbance	1.001	0.977	1.025	0.950
Daytime somnolence	1.011	0.985	1.038	0.415
Sleep adequacy	0.994	0.976	1.012	0.482
Snoring	1.011	0.993	1.029	0.225
Awaken short	1.014	0.989	1.039	0.284
Quantity of sleep	1.054	0.758	1.466	0.755

95% CI, 95% confidence interval; OR, odds ratio; MIDAS, migraine disability assessment; SF-12, 12-Item Short Form Health Survey; PCS-12, physical component summary of the SF-12; MCS-12, mental component summary of the SF-12; and NDI, neck disability index.

Table 3. Univariate linear regression to analyze the factors related to the degree of deviation of the perceived vertical from the theoretical vertical.

Variable	B	95% CI		p-Value
		Lower	Upper	
Gender	−0.026	−0.314	0.262	0.859
Smoking	0.272	−0.161	0.704	0.217
Physical activity	0.021	−0.274	0.317	0.887
Healthy/headache	−0.140	−0.425	0.145	0.334
Headache frequency	−0.005	−0.025	0.016	0.652
Headache intensity	0.003	−0.053	0.059	0.924
MIDAS	0.019	0.000	0.039	0.054
NDI	0.002	−0.017	0.020	0.868
Sleep disturbance	0.004	−0.005	0.014	0.362
Daytime somnolence	0.005	−0.006	0.015	0.405
Sleep adequacy	−0.001	−0.009	0.006	0.730
Snoring	0.004	−0.004	0.012	0.305
Awaken short	0.006	−0.006	0.017	0.323
Quantity of sleep	−0.020	−0.152	0.112	0.769

95% CI, 95% confidence interval; B, Regression coefficient; MIDAS, migraine disability assessment; F, frequency; SF-12, 12-Item Short Form Health Survey; PCS-12, physical component summary of the SF-12; MCS-12, mental component summary of the SF-12; and NDI, neck disability index.

4. Discussion

This work aimed to analyze the differences in the visual perception of verticality between healthy subjects, migraineurs, and those with TTHs using a new mobile device in conjunction with virtual reality glasses. The test was carried out without any great difficulties, and the results were stored within the developed application. The duration of each test was no more than 5 min, including the placement of the device, familiarization with it, and the performance of all required attempts.

During the development of this research, it was observed that contribution of the vestibular system to the perception of verticality remained stable in young students who present this pathology. No differences were found in the perception of verticality between healthy subjects, subjects with TTHs, and subjects with migraines. It was also observed that alterations in the perception of verticality as well as in the degrees of deviation from the perceived vertical are not related to a higher level of disability associated with neck pain; a greater frequency, intensity or disability associated with headache; or a worse quality of sleep.

Contrary to what was expected, our results showed a greater alteration of SVV in patients with migraines than in healthy controls, although without significant differences. This apparently contradictory result is in consonance with the findings of Ashish et al., 2017 [20] and Chang et al., 2019 [21] but is inconsistent with the findings of the study by Asai et al., 2009 [23]. It should be noted that, in these studies, the same measure of SVV was used as in ours (mean absolute error). The main difference between these studies is whether the head is fixed during the test. While in the study carry out by Asai et al. [23], the head was fixed at 0° during the test, in the studies conducted by Ashish et al. [20] and Chang et al. [21], the head was not fixed. Consequently, it seems that, when subject performs the test eliminating individual cervical adjustments by head fixation, the magnitude of VV deviation is greater in migraineurs than in healthy controls. However, allowing slight cervical proprioceptive adjustments during the test enables good perception of the visual vertical [20,21]. This fact suggests that, in patients with primary headache disorder, cervical afferences could act as a compensation mechanism that allows good perception of verticality. This highlights the important role played by the upper cervical structures both in the perception of verticality and in headaches. In future studies, VV measurements should be performed under different conditions and by taking into account the magnitude (absolute value) and laterality of the deviation, which would help to clarify the importance of cervical afferents and reflexes in the pathophysiology of the migraine.

Structural disorders of the upper cervical region are an important component in the pathophysiology of headaches [15,37–39]. In addition, headache and vestibular problems frequently occur together, giving rise to nonspecific balance disturbances concomitant to headache disorders [16,40,41]. These factors, in addition to the enormous importance of the information provided by these systems to shape the sense of verticality, are reasons why it is pertinent to look for a possible alteration in the perception of verticality in subjects complaining of headaches.

The most commonly used test to measure alteration of the visual perception of verticality is a static SVV test, which was used in this study. The most widespread interpretation is that this test mainly measures the contribution of the vestibular system to the perception of verticality [42]. In this sense, our results could be interpreted as a lack of a relationship between the alteration of the vestibular system and the appearance of headaches and migraines. Previously, other authors have evaluated SVV in similar condition to us. Although both their evaluation method and the number of measurements as well as the initial line position were different to those carried out in our study, their results are in line with our results, where no differences were found in the vestibular contribution to the perception of verticality between healthy subjects and subjects with headache disorders [20,43,44].

Another means of measuring visual verticality is the rod and frame test (RFT). In this test, a rod is displayed in darkness inside a tilted or untilted frame with respect to the earth vertical [45]. It has been suggested that the RFT measures more specifically the contribution of visual information and neck proprioception to the sensory integration of verticality [46]. Our results did not show a relationship of the vestibular contribution to the perception of verticality with headaches and migraines. Given that the RFT more specifically measures the contribution of visual and proprioceptive signals to the perception of verticality, we believe that it should be evaluated if differences are found between healthy subjects, migraineurs, and those with TTHs using the RFT as a measure of visual verticality.

Given that the internal model of space and verticality is constantly updated [47,48], it is speculated that, in these disorders, verticality alterations appear during the attack of headaches, reaching a balance during the interictal phase of the process [20]. This, together with the fact that measurements were conducted during a period in which the subjects were headache-free, could have conditioned the results obtained in our study.

There are several limitations of the present study. First, the population in which the study was carried out is very specific, which makes it difficult to extrapolate the results to populations with different characteristics. Another limitation is the difficulty of conducting these measurements at the time of the attack, restricting us to performing them during the phase in which the subjects were headache-free; this may have conditioned our results. Additionally, future studies should effectively measure the use of medication to control headache and whether this could affect verticality perception.

5. Conclusions

In our work, no significant differences were found in the vestibular contribution to the perception of verticality between healthy subjects or those suffering from migraines or TTHs. The variables usually related to headaches could not explain either the presence of a poor perception of verticality or the MAE during the SVV test.

For future research, it would be interesting to observe whether there are alterations in the perception of verticality during attacks as well as to observe the contribution of somatosensory and visual inputs to the perception of verticality in this and other populations. For this, it would be relevant to measure the differences in the perception of verticality between healthy and headache subjects using tests other than static SVV.

Given the versatility of new mobile devices for measuring verticality, different tests must be implemented to be able to use them both in clinical practice and in research. This could contribute to a better understanding of the pathophysiological mechanisms that are present in patients with headaches and migraines.

Author Contributions: Conceptualization, R.L.-V. and A.A.-O.; methodology, R.L.-V., D.R.-A., and A.A.-O.; software, R.L.-V. and D.R.-A.; validation, D.R.-A. and A.A.-O.; formal analysis, R.L.-V., E.O.-G., and D.R.-A.; investigation, A.A.-O., M.C.O.-P., and N.Z.-A.; resources, A.A.-O., M.C.O.-P., and N.Z.-A.; data curation, D.R.-A., M.C.O.-P., and N.Z.-A.; writing—original draft preparation, N.Z.-A. and D.R.-A.; writing—review and editing, E.O.-G. and A.A.-O.; visualization, R.L.-V. and E.O.-G.; supervision, R.L.-V. and A.A.-O.; project administration, R.L.-V. and A.A.-O.; funding acquisition, R.L.-V. and M.C.O.-P. All authors read and agreed to the published version of the manuscript.

Funding: E.O.-G. received external funding by grant number FPU17/01619 of the Ministry of Science, Innovation, and Universities, Government of Spain.

Acknowledgments: The authors thank Fidel Hita for his contribution to the diagnosis and classification of the participants in this study.

Conflicts of Interest: The authors declare that Daniel Rodríguez-Almagro and Rafael Lomas-Vega each own 14% of the intellectual property rights of the software developed for the operation of this mobile device. The funders had no role in the design of the study; in the collection, analyses, or interpretation of data; in the writing of the manuscript, or in the decision to publish the results.

References

1. Bergmann, J.; Bardins, S.; Prawitz, C.; Keywan, A.; MacNeilage, P.; Jahn, K. Perception of postural verticality in roll and pitch while sitting and standing in healthy subjects. *Neurosci. Lett.* **2020**, *730*, 135055. [CrossRef]

2. Kheradmand, A.; Winnick, A. Perception of Upright: Multisensory Convergence and the Role of Temporo-Parietal Cortex. *Front. Neurol.* **2017**, *8*, 552. [CrossRef] [PubMed]

3. Mazibrada, G.; Tariq, S.; Pérennou, D.; Gresty, M.; Greenwood, R.; Bronstein, A.M. The peripheral nervous system and the perception of verticality. *Gait Posture* **2008**, *27*, 202–208. [CrossRef] [PubMed]

4. Selge, C.; Schoeberl, F.; Bergmann, J.; Kreuzpointner, A.; Bardins, S.; Schepermann, A.; Schniepp, R.; Koenig, E.; Mueller, F.; Brandt, T.; et al. Subjective body vertical: A promising diagnostic tool in idiopathic normal pressure hydrocephalus? *J. Neurol.* **2016**, *263*, 1819–1827. [CrossRef] [PubMed]

5. Piscicelli, C.; Pérennou, D. Visual verticality perception after stroke: A systematic review of methodological approaches and suggestions for standardization. *Ann. Phys. Rehabil. Med.* **2017**, *60*, 208–216. [CrossRef] [PubMed]

6. Piscicelli, C.; Nadeau, S.; Barra, J.; Perennou, D. Assessing the visual vertical: How many trials are required? *BMC Neurol.* **2015**, *15*, 215. [CrossRef] [PubMed]

7. Dai, T.; Kurien, G.; Lin, V.Y. Mobile phone app Vs bucket test as a subjective visual vertical test: A validation study. *J. Otolaryngol. Head Neck Surg.* **2020**, *49*, 6. [CrossRef]

8. Dieterich, M.; Brandt, T. Perception of Verticality and Vestibular Disorders of Balance and Falls. *Front. Neurol.* **2019**, *10*, 172. [CrossRef]

9. Molina, F.; Lomas-Vega, R.; Obrero-Gaitán, E.; Rus, A.; Almagro, D.R.; Del-Pino-Casado, R. Misperception of the subjective visual vertical in neurological patients with or without stroke: A meta-analysis. *NeuroRehabilitation* **2019**, *44*, 379–388. [CrossRef]

10. Obrero-Gaitán, E.; Molina, F.; Del-Pino-Casado, R.; Ibáñez-Vera, A.J.; Rodríguez-Almagro, D.; Lomas-Vega, R. Visual Verticality Perception in Spinal Diseases: A Systematic Review and Meta-Analysis. *J. Clin. Med.* **2020**, *9*, 1725. [CrossRef]

11. Grabherr, L.; Cuffel, C.; Guyot, J.P.; Mast, F.W. Mental transformation abilities in patients with unilateral and bilateral vestibular loss. *Exp. Brain Res.* **2011**, *209*, 205–214. [CrossRef] [PubMed]

12. Hong, S.M.; Yeo, S.G.; Byun, J.Y.; Park, M.S.; Park, C.H.; Lee, J.H. Subjective visual vertical during eccentric rotation in patients with vestibular neuritis. *Eur. Arch. Oto-Rhino-Laryngol.* **2010**, *267*, 357–361. [CrossRef] [PubMed]

13. Zwergal, A.; Rettinger, N.; Frenzel, C.; Dieterich, M.; Brandt, T.; Strupp, M. A bucket of static vestibular function. *Neurology* **2009**, *72*, 1689–1692. [CrossRef]

14. Negrillo-Cardenas, J.; Rueda-Ruiz, A.J.; Ogayar-Anguita, C.J.; Lomas-Vega, R.; Segura-Sanchez, R.J. A System for the Measurement of the Subjective Visual Vertical using a Virtual Reality Device. *J. Med. Syst.* **2018**, *42*, 124. [CrossRef]

15. Ashina, S.; Bendtsen, L.; Lyngberg, A.C.; Lipton, R.B.; Hajiyeva, N.; Jensen, R. Prevalence of neck pain in migraine and tension-type headache: A population study. *Cephalalgia* **2015**, *35*, 211–219. [CrossRef] [PubMed]

16. Teggi, R.; Manfrin, M.; Balzanelli, C.; Gatti, O.; Mura, F.; Quaglieri, S.; Pilolli, F.; Redaelli de Zinis, L.O.; Benazzo, M.; Bussi, M. Point prevalence of vertigo and dizziness in a sample of 2672 subjects and correlation with headaches. *Acta Otorhinolaryngol. Ital.* **2016**, *36*, 215–219. [CrossRef]

17. Cachinero-Torre, A.; Díaz-Pulido, B.; Asúnsolo-Del-Barco, Á. Relationship of the Lateral Rectus Muscle, the Supraorbital Nerve, and Binocular Coordination with Episodic Tension-Type Headaches Frequently Associated with Visual Effort. *Pain Med.* **2017**, *18*, 969–979. [CrossRef]

18. Stovner, L.J.; Nichols, E.; Steiner, T.J.; Abd-Allah, F.; Abdelalim, A.; Al-Raddadi, R.M.; Ansha, M.G.; Barac, A.; Bensenor, I.M.; Doan, L.P.; et al. Global, regional, and national burden of migraine and tension-type headache, 1990–2016: A systematic analysis for the Global Burden of Disease Study 2016. *Lancet Neurol.* **2018**, *17*, 954–976. [CrossRef]

19. Woldeamanuel, Y.W.; Cowan, R.P. Migraine affects 1 in 10 people worldwide featuring recent rise: A systematic review and meta-analysis of community-based studies involving 6 million participants. *J. Neurol. Sci.* **2017**, *372*, 307–315. [CrossRef]

20. Ashish, G.; Augustine, A.M.; Tyagi, A.K.; Lepcha, A.; Balraj, A. Subjective Visual Vertical and Horizontal in Vestibular Migraine. *J. Int. Adv. Otol.* **2017**, *13*, 254–258. [CrossRef]

21. Chang, T.-P.; Winnick, A.A.; Hsu, Y.-C.; Sung, P.-Y.; Schubert, M.C. The bucket test differentiates patients with MRI confirmed brainstem/cerebellar lesions from patients having migraine and dizziness alone. *BMC Neurol.* **2019**, *19*, 219. [CrossRef] [PubMed]

22. Winnick, A.; Sadeghpour, S.; Otero-Millan, J.; Chang, T.-P.; Kheradmand, A. Errors of Upright Perception in Patients With Vestibular Migraine. *Front. Neurol.* **2018**, *9*, 892. [CrossRef] [PubMed]

23. Asai, M.; Aoki, M.; Hayashi, H.; Yamada, N.; Mizuta, K.; Ito, Y. Subclinical deviation of the subjective visual vertical in patients affected by a primary headache. *Acta Otolaryngol.* **2009**, *129*, 30–35. [CrossRef] [PubMed]

24. Kandemir, A.; Çelebisoy, N.; Köse, T. Perception of verticality in patients with primary headache disorders. *J. Int. Adv. Otol.* **2014**, *10*, 138–143. [CrossRef]

25. Martin, T.; Gauthier, A.; Ying, Z.; Benguigui, N.; Moussay, S.; Bulla, J.; Davenne, D.; Bessot, N. Effect of sleep deprivation on diurnal variation of vertical perception and postural control. *J. Appl. Physiol.* **2018**, *125*, 167–174. [CrossRef]

26. Haynes, W.; Waddington, G.; Adams, R.; Isableu, B. Relationships Between Accuracy in Predicting Direction of Gravitational Vertical and Academic Performance and Physical Fitness in Schoolchildren. *Front. Psychol.* **2018**, *9*, 1528. [CrossRef]

27. Ferini-Strambi, L.; Galbiati, A.; Combi, R. Sleep disorder-related headaches. *Neurol. Sci.* **2019**, *40*, 107–113. [CrossRef]

28. Rodríguez-Almagro, D.; Achalandabaso-Ochoa, A.; Obrero-Gaitán, E.; Osuna-Pérez, M.C.; Ibáñez-Vera, A.J.; Lomas-Vega, R. Sleep Alterations in Female College Students with Migraines. *Int. J. Environ. Res. Public Health* **2020**, *17*, 5456. [CrossRef]

29. Amin, F.M.; Aristeidou, S.; Baraldi, C.; Czapinska-Ciepiela, E.K.; Ariadni, D.D.; Di Lenola, D.; Fenech, C.; Kampouris, K.; Karagiorgis, G.; Braschinsky, M.; et al. The association between migraine and physical exercise. *J. Headache Pain* **2018**, *19*, 83. [CrossRef]

30. Vandenbroucke, J.P.; von Elm, E.; Altman, D.G.; Gøtzsche, P.C.; Mulrow, C.D.; Pocock, S.J.; Poole, C.; Schlesselman, J.J.; Egger, M. Strengthening the Reporting of Observational Studies in Epidemiology (STROBE): Explanation and elaboration. *Int. J. Surg.* **2014**, *12*, 1500–1524. [CrossRef]

31. Wöber-Bingöl, C. Epidemiology of migraine and headache in children and adolescents. *Curr. Pain Headache Rep.* **2013**, *17*, 341. [CrossRef] [PubMed]

32. Olesen, J.; Bes, A.; Kunkel, R.; Lance, J.W.; Nappi, G.; Pfaffenrath, V.; Clifford Rose, F.; Schoenberg, B.S.; Soyka, D.; Welch, K.M.A.; et al. The International Classification of Headache Disorders, 3rd edition (beta version). *Cephalalgia* **2013**, *33*, 629–808. [CrossRef] [PubMed]

33. Rodríguez-Almagro, D.; Achalandabaso, A.; Rus, A.; Obrero-Gaitán, E.; Zagalaz-Anula, N.; Lomas-Vega, R. Validation of the Spanish version of the migraine disability assessment questionnaire (MIDAS) in university students with migraine. *BMC Neurol.* **2020**, *20*, 67. [CrossRef] [PubMed]

34. Andrade Ortega, J.A.; Delgado Martínez, A.D.; Almécija Ruiz, R. Validation of a Spanish version of the Neck Disability Index. *Med. Clin.* **2008**, *130*, 85–89. [CrossRef] [PubMed]

35. Zouabi, A.; Quarck, G.; Martin, T.; Grespinet, M.; Gauthier, A. Is there a circadian rhythm of postural control and perception of the vertical? *Chronobiol. Int.* **2016**, *33*, 1320–1330. [CrossRef]

36. Zagalaz-Anula, N.; Hita-Contreras, F.; Martínez-Amat, A.; Cruz-Díaz, D.; Lomas-Vega, R. Psychometric properties of the medical outcomes study sleep scale in Spanish postmenopausal women. *Menopause* **2017**, *24*, 824–831. [CrossRef]

37. Ford, S.; Calhoun, A.; Kahn, K.; Mann, J.; Finkel, A. Predictors of disability in migraineurs referred to a tertiary clinic: Neck pain, headache characteristics, and coping behaviors. *Headache* **2008**, *48*, 523–528. [CrossRef] [PubMed]

38. Carvalho, G.F.; Chaves, T.C.; Gonçalves, M.C.; Florencio, L.L.; Braz, C.A.; Dach, F.; Fernández de Las Peñas, C.; Bevilaqua-Grossi, D. Comparison between neck pain disability and cervical range of motion in patients with episodic and chronic migraine: A cross-sectional study. *J. Manip. Physiol. Ther.* **2014**, *37*, 641–646. [CrossRef]

39. Blaschek, A.; Milde-Busch, A.; Straube, A.; Schankin, C.; Langhagen, T.; Jahn, K.; Schröder, S.A.; Reiter, K.; von Kries, R.; Heinen, F. Self-reported muscle pain in adolescents with migraine and tension-type headache. *Cephalalgia* **2012**, *32*, 241–249. [CrossRef]

40. Swaminathan, A.; Smith, J.H. Migraine and vertigo. *Curr. Neurol. Neurosci. Rep.* **2015**, *15*, 515. [CrossRef]

41. Bisdorff, A.; Andrée, C.; Vaillant, M.; Sándor, P.S. Headache-associated dizziness in a headache population: Prevalence and impact. *Cephalalgia* **2010**, *30*, 815–820. [CrossRef] [PubMed]

42. Perennou, D.; Piscicelli, C.; Barbieri, G.; Jaeger, M.; Marquer, A.; Barra, J. Measuring verticality perception after stroke: Why and how? *Neurophysiol. Clin.* **2014**, *44*, 25–32. [CrossRef] [PubMed]

43. Miller, M.A.; Crane, B.T. Static and dynamic visual vertical perception in subjects with migraine and vestibular migraine. *World J. Otorhinolaryngol. Head Neck Surg.* **2016**, *2*, 175–180. [CrossRef] [PubMed]

44. Crevits, L.; Vanacker, L.; Verraes, A. Patients with migraine correctly estimate the visual verticality. *Clin. Neurol. Neurosurg.* **2012**, *114*, 313–315. [CrossRef] [PubMed]

45. Bagust, J.; Docherty, S.; Haynes, W.; Telford, R.; Isableu, B. Changes in rod and frame test scores recorded in schoolchildren during development–a longitudinal study. *PLoS ONE* **2013**, *8*, e65321. [CrossRef]

46. Humphreys, B.K. Cervical outcome measures: Testing for postural stability and balance. *J. Manip. Physiol. Ther.* **2008**, *31*, 540–546. [CrossRef]

47. Glasauer, S.; Dieterich, M.; Brandt, T. Neuronal network-based mathematical modeling of perceived verticality in acute unilateral vestibular lesions: From nerve to thalamus and cortex. *J. Neurol.* **2018**, *265*, 101–112. [CrossRef]

48. Barra, J.; Perennou, D. Is the sense of verticality vestibular? *Neurophysiol. Clin.* **2013**, *43*, 197–204. [CrossRef]

MDPI

St. Alban-Anlage 66

4052 Basel

Switzerland

Tel. +41 61 683 77 34

Fax +41 61 302 89 18

www.mdpi.com

Diagnostics Editorial Office

E-mail: diagnostics@mdpi.com

www.mdpi.com/journal/diagnostics